圖解

點心製作教程

An Illustrated Guide to
Dim Sum

蔡潔儀 編著

萬里機構・飲食天地出版社

U0063819

前言 Preface

　　轉眼間教授烹飪已有二十多年，回想起 1989 年 4 月 11 日，當天情景還歷歷在目，那可算是我人生中最難忘的一天，當日下午 2 時 30 分，我正式投身烹飪行業，展開人生新的一頁。

　　能以自己的興趣作為終生事業，除了要感謝上天的恩賜外，最使我銘感於心的，是一直以來給予我莫大支持和鼓勵的各界前輩和烹飪學生們。我深切盼望，能在有生之年，將我所學所得的烹飪知識，盡量留給我的學生和與我一樣熱愛烹飪的人士，以回謝各位對我的厚愛。

　　在芸芸食品中，甜品、點心和小食，是我最喜愛的一環，這亦是許多人心中的美食。製作點心，就如學做人一樣，不能草率輕浮、得過且過。每次當我拿起材料製作時，都有一股推動力不停的自我反省。這也正是我喜愛做點心的原因之一。

　　本書寫作方法較一般的烹飪書更為詳盡。由材料開始至完成的每一個環節，都有清晰講解和流程圖，能使讀者清楚瞭解和掌握箇中竅門，是一本讓讀者容易學習的書籍。

　　一款好的點心，除了入口好吃外、精緻的造型更是吸引人的魅力所在。一般而言，學習製作點心比烹調其他家常菜式困難。一般菜式大多可自由加減調配，鹹、淡、脆、軟，悉隨尊便，縱然比例稍有偏差，亦可將之變化為創新佳餚，無傷大雅。做點心卻不然，必須留意的事項十分多，如材料的份量、比例、濕度、氣溫變化以至揉捏的力度等等……。草率了事，可能難以成型及破壞效果。此外，最重要的還是必須配合製作者的手藝、熱誠和耐性，多學多做、屢敗屢試，才能使其盡善盡美。

蔡潔儀

Before I come to aware of it, I have been teaching cooking over 20 years. Looking back on the day of April 11, 1989, at 2:30 in the afternoon, I know it is the most unforgettable time in my life: the start of my cooking career.

Being able to make personal interest as a life career, other than thank god's care for me, I would hold most appreciation to personage from various circles and my cooking students, who have been giving me great support all the way up. Deep in my heart, in return for their kindness, I hope in my life time, I can pass the cooking knowledge I've got as much as possible to my students and those who loving cooking.

Among all sorts of courses, dessert, pastry and snack are my favorites. They are also the most delicious stuffs for many people. Making dessert is like conducting oneself in the society: you can't be rash and frivolous; you can't muddle things along. Every time I take up the ingredients, there is always a force driving me to do some self-examination. This is also the reason why I love making dessert.

This book of is written with the aim to make it a cooking textbook. For every steps from preparing ingredients to finishing, there are detailed explanations and flow charts, which helps readers clearly understand and grasp all the necessary keys. In a word, this is a cooking book easy to learn.

Besides the taste, delicate mould is where the charm comes from a good dessert. In my own opinion, learning to make dessert is more difficult than learning to cook other daily dishes. The ingredients, the saltiness, the lightness, the crispiness and the softness of general dishes mostly can be adjusted freely to your taste. The differences from the standard ones don't matter much and can be made the base for further created dishes. But it's not the case for making dessert. There are many to notice, e.g., the amount and portion of ingredients, the humidity, the change of temperature and the strength used to knead dough, etc. If you deal with these at your please, it is very possible that you can't make it in form or destroy the effect. In addition, the maker's skills, enthusiasm and patience are important. And the key to make your dessert as perfect as possible is to learn more and do it yourself more; fail many times and try more times again.

Kitty Choi

目錄

中難度
Intermediate Pratice　　　　　　72

用具與點心的基礎認識
Basic Knowledge for Dim Sum and Tools

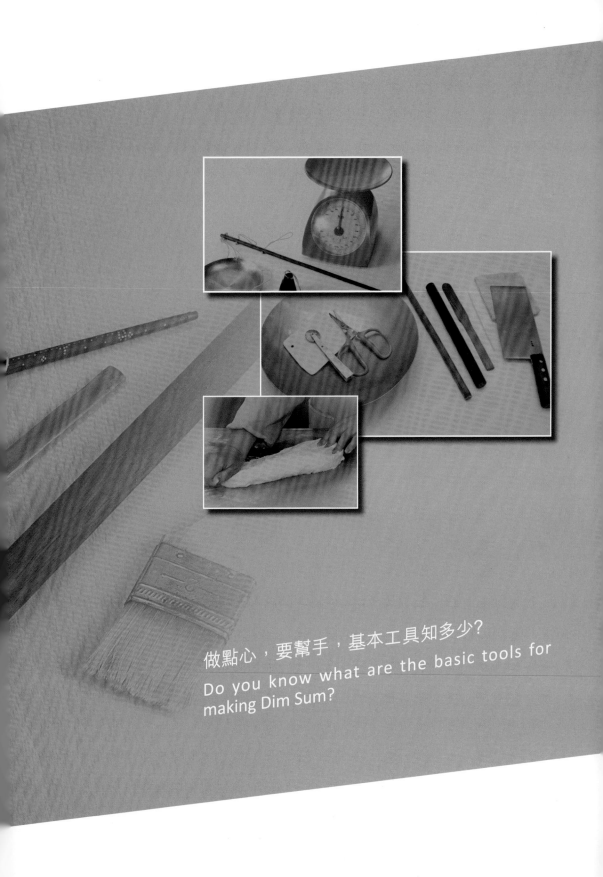

做點心，要幫手，基本工具知多少？

Do you know what are the basic tools for making Dim Sum?

❶ 中國秤：量度微小份量的材料，一般量度單位是分、錢、兩。
Chinese scale is for measuring small amounts of ingredients; measuring units generally include candareen, mace and tael.

❷ 磅：量度大份量的材料，一般量度單位是安士、克等。
Dial scale is for measuring larger amounts of ingredients; measuring units generally include ounce, gram, etc.

❸ 刮刀(不鏽鋼)：融合材料、分割麵糰、刮刀(塑膠)。
Dough scraper (stainless steel) is for mixing ingredients and dividing dough. Dough scraper (plastic).

❹ 餡挑／餡料撥刀：又稱餡刮皮，以竹片製成，上餡之用。
Filling spatula / filling blade, also called filling scraper, is made of bamboo. It is used for stuffing.

❺ 酥棍：比點心棍略粗和短一點，光滑，用於開小酥或麵皮之用。
Pastry rolling pin is a little bit thicker and shorter than the dim sum rolling pin. It is smooth and is used for making small pastry or dough wrapper.

❻ 油掃：餃子類成品需要在成熟前掃油，或在包點上掃蛋液之用。
Oil brush is used for brushing oil on dumplings before they are done or brushing egg mixture on buns.

⑦ 壓紋器：在成品壓紋裝飾。
Pattern press is for decorating products with pressed patterns.

⑧ 剪刀：弄斷材料和修改製品。
Scissors are for cutting ingredients and trimming products.

⑨ 撈餡碟：多為金屬製或陶瓷，用於混合或拌和材料、盛載餡料。
Filling-mixing plate is usually metal-made or ceramic-made. It is used for mixing or stirring ingredients and holding filling.

⑩ 輾麵機：麵糰搓揉後輾薄成麵皮的機器。
Pasta machine is a machine for flattening out kneaded dough into a thin layer.

⑪ 點心棍 / 單手杖 / 小麵杖：長約24厘米，光滑纖細，檀木材質不易變形，用於擀餃子。
Dim sum rolling pin, about 24 cm long, smooth and subtle, is used for flattening dumpling wrapper. Being made of sandalwood, it does not change shape easily.

⑫ 油布 / 濕布：它是開皮刀的輔助工具，抹刀後能防止刀身黏刀。
Oil towel or wet towel is the assisting tool for pastry knife. The blade will not be sticky after wiping the knife with the towel.

⑬ 開皮刀：外形如菜刀，沒有開鋒不銳利，行內稱拍皮刀，以片蝦餃皮為主。
Unsharpen pastry knife, which looks like a Chinese chopper, is blunt. Also called dim sum knife for insiders, it is mainly used for slicing shrimp dumpling wrappers.

泡打粉 / 發粉：混合了酸鹼的化學膨脹劑，每當與麵糰或麵糊之類存有濕氣的物質接觸，就會引起化學反應，產生氣體令麵糰或麵糊內有微細氣泡的作用。

Baking powder is a chemical raising agent mixed with acid and alkali. When it is in contact with moist substances such as dough and batter, chemical reactions occur, producing gas which makes the dough or batter have small air bubbles.

吉士粉：又稱雞蛋粉，色澤微黃，與水融合變淡黃色，含獨特香味，由穩定劑、食用香油、食用色素、奶粉和粟粉等組合而成。

Custard powder, also called egg powder, is slightly yellow and becomes pale yellow when it is mixed with water. It has a unique fragrance and is composed of stabiliser, edible fragrance oil, edible pigment, milk powder and corn starch.

麵粉：含灰質少於1.25%，麵筋含量不低於22%，而水份含量就不超過12.5%，韌度低，適合做一般無韌度要求的麵點。

Plain flour contains gluten less than 1.25%, gluten not less than 22% and water not more than 12.5%. It is of low elasticity and is suitable for making products with no elasticity requirement.

粟粉：由完全成熟的粟米抽取而成的澱粉，含二氧化硫、色潔白，粉粒微細，觸感輕軟，能令成品變幼滑細緻，輕身綿密。

Corn starch is the starch extracted from fully mature maize. It contains sulphur dioxide. It is white and fine with a light and soft touch. It can make the products smooth and subtle, and light and dense.

臭粉：這是一種有味的麵粉添加劑，呈幼粉末狀，含阿摩尼亞的味道，十分嗆鼻，其作用是令成品出現爆裂現象，兼有延伸貯藏時間。

Ammonium bicarbonate is a smelly flour additive. It is fine powder with the smell of ammonia and is very pungent. Its functions are to make the products crack and to prolong storage time.

中筋粉：含麩質不超過1.25％，麵筋質含量不低於24％，水份含量不超過14％，色澤潔白，韌度中等，粉質幼細，成品疏鬆有光澤，適合做包類和酥皮。

All-purpose flour contains gluten not more than 1.25%, gluten not less than 24% and water not more than 14%. It is white and fine with medium elasticity. It is suitable for making buns and pastries. The products made are spongy and shiny.

粘米粉：用米研磨而成粉末，含穀蛋白和穀黏膠蛋白，遇水後不能組成麵筋，麵糰沒有筋性，以做糕為主。

Rice flour is made by grinding rice into powder. It contains glutenin and gliadin, and does not form gluten when it is added with water. It has no elasticity and is mainly used for making cakes.

澄麵：原屬小麥粉的產物，當小麥研磨後，與清水混和，再經過過濾沉澱，浮在最上層的粉末，取出乾燥而成。它的筋性低，粉末幼細，色潔白，無味，硬度足，常與生粉一同使用。

Wheat starch (Tang flour) is a product of wheat powder. When wheat is ground and mixed with water, and then filtered and settled, there is powder floating on the top. The powder is collected and dried, which is the wheat starch. Wheat starch is of low elasticity. It is fine and white with no taste. It is hard and is usually used with tapioca starch.

馬蹄粉：由馬蹄研磨而成的重澱粉，顆粒粗，含有一股清甜味道，當與水結合，烹熟後黏度高，呈半透明，韌度高而富彈力。

Water chestnut flour is a heavy starch made by grinding water chestnut. It has coarse grains and a refreshing sweet flavour. When it is mixed with water and brought to the boil, it is sticky and translucent with high pliability (yet tough) and elasticity.

糯米粉：用糯米研磨而成粉末，因含支鏈澱粉，故最易因受熱變性而糊化。

Glutinous rice flour is made by grinding glutinous rice into powder. As it contains amylopectin (a highly branched starch molecule), it is the most easily denatured due to high temperatures and becomes paste.

13

食用色素：分有天然和人造兩大類，以調色為主，增加成品的顏色。
There are two types of edible pigments: natural and artificial. They are used for colouring to increase the colour of the products.

砂糖：原材料是蔗糖或甜菜頭提煉而成，精煉後去除糖蜜部份，變成沒有糖香的白晶體。
Granulated sugar is refined from cane sugar or beet. After refining, treacle is removed and the sugar becomes white crystal without sweet smell.

牛油：乳製加工品，分有機和普通牛油，尤以丹麥和法國的品質最好，至於澳洲和紐西蘭的牛油，味道濃郁兼有羶味。
Butter is processed dairy product. There are organic butter and common butter. Butter from Denmark and France has the best quality while butter from Australia and New Zealand has rich flavour but with a smell of mutton.

糖霜：白糖研磨成細碎粉末狀，並混入少許粟粉，防止黏貼融合，它能迅速溶解。
Icing sugar is made by grinding white sugar into fine powder and mixing the powder with a little corn starch to prevent it from clumping. It can dissolve rapidly.

酵母 / 依士：一種單細胞微生物活菌，當它開始活動繁殖，就會產生發酵作用，然後將食物如糖，配合適當溫度35℃ - 42℃，繼而轉換成酒精和二氧化碳，當中的二氧化碳會令麵包或包子膨脹起來。
Yeast is a single-celled microorganism. Under the temperature ranging from 35°C to 42°C, when it is active and reproduces, fermentation occurs, changing foods such as sugar to ethanol and carbon dioxide. The carbon dioxide produced makes the bread or buns expand.

豬皮：屬豬的外皮，含天然骨膠原，去掉皮下脂肪，經長時間烹煮，就能產生膠質的液體，冷凍後自動凝固，加熱後轉態成液體。

Pig skin is the outer skin of pig. It contains natural collagen; gelatinous liquid can be made by removing the subcutaneous fat from the pig skin and cooking the pig skin for a long period of time. The gelatinous liquid can be solidified by cooling and returns to liquid upon heating.

雞腳：它是湯凍的主要功臣，採用黃油雞腳，肉厚味道濃郁，熬湯後會呈現金黃油層，色澤艷麗，與豬皮有異曲同功之妙。

Chicken feet are the main ingredient for preparing stock gel. Using yellow-oil chickens' feet which is fleshy provides a rich flavour. A golden layer of oil forms when making soup from them. The soup is brightly coloured and has the same function as the pig skin.

雞蛋：味道香濃，可塑成高，用途廣泛，調色、調味和令成品膨脹。

Eggs are of rich flavour and with great potential. They are versatile, such as for colouring, seasonings and making the products expand.

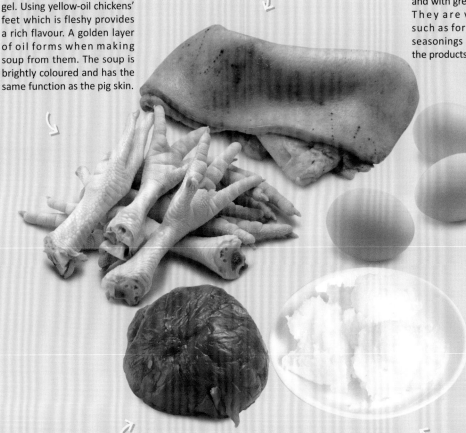

蓮蓉：由蓮子烹熟磨蓉，與糖和油炒香而成常用餡料。

Lotus paste is made by cooking lotus nuts until done and grinding the lotus nuts into paste. It is stir-fried with sugar and oil until fragrant to become a filling, which is commonly used.

豬油：以豬膏（豬大油）或肥豬肉精製，繼而進行淨澄清過程，造到脫臭和脫色的動物油脂。

Lard is made from pig fat or fatty pork. After cleaning and purifying processes, decolourised animal fat free from smell is obtained.

麵包製作材料過程與認識
Ingredients and Processes for Bread Making

麵包的製作材料：麵粉

麵粉分高筋(高蛋白)、中筋(中蛋白)及低筋(低蛋白)三種：

高筋：
顏色較黃，本身較有活 且光滑，含蛋白質13.5%，吸水力為60-64%，適用於一般麵包。

低筋：
顏色較白，俗稱白麵粉，含蛋白質8.5%，吸水力為50-53%，適用於一般蛋糕。

中筋：
介於高筋、低筋之間，顏色乳白，體質半鬆性，含蛋白質11%，吸水力為55-58%，適用於中國點心。

FLour for Making Bread

The three types of flour used include high gluten (high protein), medium gluten (medium protein) and low gluten (low protein) :

High Gluten:
Slightly yellow in colour, smooth and contains active ingredients. It is made up of 13.5% protein and water absorption is 60-64%. Suitable for most bread types.

Low Gluten:
Whiter in colour and is commonly known as white flour. Contains 8.5% protein and water absorption up to 50-53%. Suitable for most cakes.

Medium Gluten:
Between high and low gluten. Pure white in colour and is semi loose in texture. Contains 11% protein and water absorption of 55-58%. Suitable for making Chinese dim sums.

1

2

3

4

5

6

7

麵包的製作過程

1 **水化階段**：將麵粉篩在桌上，把配方全部材料放入，進行拌勻及搓揉。

2 **麵糰捲起及擴展階段**：搓揉至所有材料混合成膠黏糰狀，直至看不見油脂為止，繼續搓揉成三不黏。

3 **麵糰完成階段**：用手拉取麵糰時，具有良好之擴展力及彈 。

4 **發酵階段**：將麵糰放在抹油的盤中，覆蓋一塊白布，置濕暖處（溫度約26-28℃，相對濕度為75-80%），基本發酵時間約90分鐘。

5 **測試**：麵糰發酵至原來體積約兩倍大，即可用手指黏上少許麵粉，戳洞檢查發酵情況，如凹入的地方沒有升起即表示發酵完成。

6 **翻麵階段**：用拳頭壓向麵糰散氣，然後由外向內摺置麵糰進行翻麵。

7 **分割**：把翻好之麵糰分割成同一重量的等份。

8 **滾圓**：將手掌張開，然後輕輕抓住麵糰，將四指併排，指尖向內彎曲，輕微地依順時針方向轉動，麵糰自然合由長形滾成圓球狀，感覺麵糰已成圓球，光滑而結實，因不宜過久，應立即停止。滾圓好的麵糰，放在撒有少許麵粉的桌面或木板上，使其繼續發酵約30分鐘（此亦稱為中間發酵）蓋上白布，以免麵糰的表皮乾燥。

9 **造形及最後發酵**：30分鐘後將滾圓之麵糰取出，進行造形，然後放在塗有一層薄油的烤盤上作最後發酵（約30分鐘），於入爐煎塗上一層蛋液。

Steps for Making Bread

1 **Mixing with water:** Sieve flour on table and add side ingredients. Mix and knead.

2 **Gathering dough & Expanding:** Knead until all ingredients are well mixed and dough is elastic. None of the fat should be visible. Continue kneading until it reaches the three non-stick stage.

3 **Finishing the dough:** When pulling the dough, it should stretch and is elastic.

4 **Fermenting Stage:** Place dough in a greased bowl, cover with white cloth and place in a warm and wet area (26-28°C high and humidity of 75-80%). Basic rising time is about 90 minutes.

5 **Testing:** The dough should rise to double its original volume. Dust finger with flour and poke into dough to test. If the dented area does not rise, that means the fermenting is complete.

6 **Turning the Dough:** Punch dough with fist to expel air. Fold from outside in and start to turn dough over.

7 **Divide:** Cut turned dough into portions of the same weight.

8 **Rolling:** Open palm of hand and hold dough lightly. Line all four fingers in a row with tips facing inwards. Roll dough in a slight clockwise direction. The dough will gradually change from long shape to round one. It should also be firm and smooth. When you can feel when this is achieved, stop rolling immediately as the dough should not be over worked.

Place the round dough balls on a slightly floured board or table top. Let them continue to prove for 30 minutes (this is known as the middle rising stage), cover with white cloth to prevent the surface drying.

9 **Shaping & Final Proving:** Remove round dough balls after 30 minutes. Shape and place on a lightly greased baking tray for the final proving (about 30 minutes). Brush with beaten egg before placing in oven.

點心實習課

Lessons for Dim Sum

講技法，論手藝，圖文並茂，實習最重要。

家庭手作點心，邁向專業門檻，就要經過三級跳試煉，沒有捷徑可尋，只有身體力行，眼到、手到、心到，從基本功出發，按步就班，揣摩技術，不斷練習，技術成熟了，就要拓闊眼界，與生活接軌，開展創意。

做點心，考手藝，有易亦有難，為求成功，就要從簡單處着手，技術掌握了，再向高難度挑戰，成功在望。

From family-style dim sum to the professional one, three levels of training must be undergone and there is no short cut. Only through putting into practice, with your eyes, your hands and your heart, starting from the basic techniques, then step by step, digging into the techniques and practising continuously, may one be skilful. When you become skilful, you should then widen your vision and connect with life to develop your creativity.

Dim sum making emphasise skills; some are easy while some are difficult. To succeed, we need to start from the easy ones first and challenge the more difficult ones after having manipulated the skills. Then the success is just ahead.

低難度
Elementary Pratice

基礎麵點製作流程圖
Basic Production Flow Chart

原輔料 Raw and auxiliary materials	選料 Selecting ingredients	• 冷水麵糰 　cold water dough
↓	↓	• 熱水麵糰 / 熟麵糰 　hot water dough / cooked dough
和麵 mix the dough	調製麵糰 Making dough	• 發酵麵糰 　sourdough
↓	↓	• 礬鹹鹽麵糰 　fried dough
揉麵 knead the dough	製坯 Make the base	
↓	↓	
搓條 and roll the dough into strip	製皮 Make the skin / wrapper	分坯 Divide the base

成形　Form the shape　←　上餡　Stuffing　←　製餡　Make the filling

熟製　Cook until done　→　裝飾　Decorate

註：
1. 冷水麵糰含強韌度、有彈力，技法用了搗、揣、摔、重複揉搓，使麵糰吸水均勻，表面光滑，調製大麵糰時手力不夠，可以木桿或竹桿子來壓，才能使筋性揉出來。
2. 熱水麵糰剛好相反，在和麵、揉麵的過程中，邊和邊加水，同時地攪和揉搓，完成後就不再搓揉，防止產生麵筋，失掉柔軟的特點。
3. 發酵麵糰在和麵時，力度要適中，不能用力過大，務求搓揉均勻透徹。待加入鹼時就要用揣的動作，使鹼水能分佈均在麵糰中。
4. 礬鹹鹽麵糰在和麵時，就要重複搗、揣、摺疊、錫麵，目的是讓麵糰達到既能膨脹，又要有韌度的要求。

Remark:
1. Cold water dough is very tough, yet pliable and elastic. The techniques involved include pounding, beating, throwing, kneading repeatedly to make the dough absorb water evenly and smooth. When making big dough, a wooden rod or bamboo rod can be used instead of bare hands to press so as to be powerful enough to bring out the elasticity of the dough.
2. Opposite for hot water dough, in the process of mixing and kneading the dough, add water while mixing, and stir and knead at the same time. When finished, do not knead to prevent gluten from forming and losing the softness characteristic.
3. When mixing sourdough, the strength applied should be suitable / medium and cannot be too large, aiming to knead the dough evenly and thoroughly. When alkali is added, beat the dough instead of kneading to make the alkaline water distribute evenly in the dough.
4. When mixing fried dough, pounding, beating, folding and lifting have to be repeated to make the dough expand and meanwhile pliable yet tough.

基本技法
Techniques and Skills

搓條

分坯／分體

製皮

上餡

搓條將麵糰搓成粗幼均勻的圓柱體，以便分坯。

Rolling strip Roll the dough into cylinders of even thickness for dividing bases.

分坯／分體可用雙手撕斷（揪劑）：手按粗大搓條，以四指伸入挖斷（挖劑）；用手指拉斷（接劑）；麵糰按平均，用刀切方塊，擀成圓形（切劑）；用刀以一刀一刀剁下（剁劑）。

Dividing base Tear apart with hands; hold the thick strip with one hand, and dig with four fingers to split; pull apart with fingers; press the dough evenly, cut into squares with knife, and flatten into circles; chop with knife piece by piece.

製皮主要用按、擀等方法，把麵坯弄成各類坯皮，配合上餡和成形。

Making skin mainly by pressing and flattening. Make the base into different kinds of skins for stuffing and shaping.

上餡將餡心放在坯皮內，緊緊包裹，不露餡。

Stuffing Put filling into skin and wrap tightly. Do not let the filling ooze out.

和麵

和麵把材料用手混和一起，變成糰。

Mixing dough Mix ingredients together with hands into dough.

揉麵

揉麵可分為搗、揉、揣、摔、擦等五個動作，能進一步使麵糰均勻，增加彈力，柔潤，光滑和軟糯。

Kneading dough There are five actions including pounding, kneading, beating, throwing and rubbing. They can further make the dough even, and increase its elasticity, pliability, smoothness and tenderness.

搗

搗和麵後把麵糰放在盆內，緊握拳頭，並在麵糰各處，用力向下搗壓，力量越大越好。

Pounding Put the dough into a basin after mixing, pound on every part of the dough with fists as powerful as possible.

揉

揉調製麵糰的重要動作，可使麵糰的澱粉膨脹帶濕潤，最終黏結，使蛋白質均勻吸水，產生彈力的麵筋網絡，增強麵糰的韌度。用雙手掌根壓實麵糰，用加推收麵糰，揉至一定程度，雙手交叉往兩旁攤開、捲疊、再攤開、捲疊，揉至麵糰變光滑。用力時以陰揉力處理，即力度要有，柔勁推開。

Kneading an important action in making dough. This makes the starch of the dough expand and become moist, and stick together finally, so that protein absorbs water evenly to form elastic gluten network, increasing the pliability and toughness of the dough. Press the dough firmly with the bottom of the palms, push and pull the dough. Knead to certain degree, then cross the hands to spread the dough onto two sides, and roll and fold, then spread again and roll and fold; knead until the dough becomes smooth. The force used should be soft force. That is there is force, but pushing out with soft force.

揣

揣緊握雙拳，交叉在麵糰上揣壓，邊揣邊壓又邊推，向外揣開，捲攏再揣，從小塊小塊進行，有時還需要沾水。

Beating Hold the fists firmly and cross to beat and press on the dough. Beat, press and push at the same time; beating outwardly, rolling and beating again. Do it bit by bit; and add water when necessary.

擦

擦用於油酥麵糰和部份米粉麵糰，做法是把油與麵和好，用手掌跟一層一層向前推擦麵糰，推擦開後滾回身前，捲攏成糰，不斷重複直至擦透。目的是令油和麵結合均勻，增強麵糰的黏度，能成品減少鬆散狀態。

Rubbing applied in pastry dough and some rice flour dough. Mix oil and flour well. Push to rub the dough layer by layer with the bottom of the palms. After spreading the dough by pushing and rubbing, roll back the dough to your side into a lump. Repeat until rub thoroughly. The aim is to mix the oil and flour evenly to increase the stickiness of the dough and avoid the products becoming loose.

按皮

按皮用右手掌面按成邊薄中間厚的圓形皮。

Pressing skin With the right palm, press the dough into a circular skin thick in the middle and thin at the edge.

拍皮

拍皮不用揉圓麵糰，用手指掀壓一個，再用手掌沿坯邊着力拍，邊拍邊順時針轉動，最後變成中央厚而四周薄的坯子。

Slapping skin No need to knead the dough into a round shape. Press one side with a finger, and then slap with the palms along the edge of the base powerfully. Turn the base clockwise while slapping. Finally, a base thick in the middle and thin at the edge is obtained.

摔

摔把麵糰舉起兩端，手不離麵，摔在案板上，摔均勻為要。另一做法，單手舉麵糰用力摔在盆中，不斷重複至所需效果。

Throwing Raise two ends of the dough, and throw the dough onto the work surface without losing hold of it. It is important to throw evenly. Another way is holding the dough with one hand, and throwing it at a basin. Repeat until the desired effect achieved.

芝麻杏仁糖
Sesame Seed Almond Candy

材料

金黃砂糖 3 3/4 兩（150克）

麥芽糖 1 1/4 兩（50克）

芝麻粉 1 1/4 兩（50克）

碎杏仁粒 1 3/4 兩（70克）

清水 1 湯匙

Ingredients

3 3/4 taels (150g) golden brown sugar

1 1/4 tael (50g) malt

1 1/4 tael (50g) sesame seed powder

1 3/4 tael (70g) chopped almond

1 tbsp water

製法

1 碎杏仁，用白鑊炒至香脆。

2 芝麻粉與糖拌勻置鑊中，加清水1湯匙，慢火熬至滴糖漿入冷水中能立即成固體顆粒狀為度，即可關火。

3 加入杏仁碎拌勻。

4 把糖料傾在保鮮紙或牛油紙上，利用保鮮紙將糖揉搓成圓柱體，在尚未完全冷卻時，將糖切成所需之厚度。

5 待糖硬後，即可以彩色玻璃紙包裹。

Method

1 Fry chopped almonds in dry pan until fragrant and crispy.

2 Mix sesame seed powder and sugar in pan. Add 1 tablespoon of water and cook on low heat until drops of the syrup go solid when dropped in cold water. Turn off heat.

3 Add chopped almonds and mix well.

4 Place sugar mixture on waxed paper or cling film and using the cling film, roll into a round column. Cut candy into desire thickness before the mixture gets completely cold.

5 Once set, pack with colour paper.

材料

純朱古力4兩（160克）

棉花糖1 3/4兩（70克）

無鹽花生1 1/2兩（60克）

顏色玻璃紙適量

Ingredients

4 taels (160g) plain chocolate

70g marshmallows

60g unsalted peanuts

Ome colour cellophane paper

石板街朱古力
Rocky Road Chocolate

製法

1 純朱古力切碎。

2 坐於溫水中至溶解。

3 將花生研成碎粒、棉花糖加入朱古力溶液中，混和後倒落盆中，雪凍。

4 切成小粒，用玻璃紙包着即成。

Method

1 Chop plain chocolate.

2 Sit over warm water until melted.

3 Grind peanuts finely and add to melted chocolate with marshmallows. Mix well and pour into pan. Chill.

4 Cut into small pieces and wrap with cellophane paper.

材料
粘米粉8兩（320克）
清水24兩（960毫升）
鹽1茶匙
乾葱（切片）4粒
蒜茸1茶匙

醃料
冬菇4隻
菜脯1兩（40克）
攪碎豬肉2兩（80克）
芫荽（切碎）1棵
葱（切粒）1條

調味料
醬油1茶匙
蠔油1湯匙
糖1/2茶匙
麻油少許
胡椒粉少許

Ingredients
320g rice flour
960ml water
1 tsp salt
4 cloves shallot (sliced)
1 tsp minced garlic

Filling
4 mushrooms
40g preserved turnip
80g minced pork
1 parsley (chopped)
1 sprig spring onion (diced)

Seasonings
1 tsp thick soy sauce
1 tbsp oyster sauce
1/2 tsp sugar
Dash of sesame oil
Pinch of pepper

製法
1　將粘米粉用清水及鹽開勻。
2　小碗排放在蒸籠內預先蒸熱2分鐘，將已攪勻之材料倒入小碗中，用猛火大滾水蒸6分鐘，待凍後起出。
3　餡料切成細粒剁碎，用油爆香蒜茸及餡料，加入調味料兜勻備用。
4　乾葱用熱油炸脆。
5　與餡料拌勻，放在碗仔糕中心位置，加上芫荽和葱粒，即可進食。

Method
1　Place rice flour in a mixing bowl, stir in water and salt to make into smooth batter.
2　Arrange several small bowls in a steamer and steam for 2 minutes. Pour the batter into the heated small bowls and cook for 6 minutes over high heat. Set aside to cool and remove the steamed cakes from the bowls.
3　Dice mushrooms and preserved turnip finely. Heat a wok with oil, add in minced garlic and filling and saute. Add seasonings to it, toss well and set aside.
4　Deep fry sliced shallots in hot oil until crispy.
5　Add in filling and toss thoroughly. Put some filling in the centre of each cake. Garnish with parsley and spring onion and serve.

材料

熱水6兩(240克)

糖3兩(120克)

豬油1湯匙

糕粉3.5兩(140克)

香蕉油1/2茶匙

紅色素1/8茶匙

Ingredients

6 taels (240g) hot water

3 taels (120g) sugar

1 tbsp lard

3.5 taels (140g) cooked
glutinous rice flour

1/2 tsp banana essence

1/8 tsp red food
colouring

製法

1　先將豬油、糖放在膠盤內,用大滾水拌溶。

2　加入香蕉油攪勻,最後加入糕粉(一邊加糕粉,一邊要用打蛋器攪勻成粉糰)。

3　先取起2兩(80克)粉糰,加入色素,搓勻備用。

4　用一張大牛油紙,上面放粉糰,並用牛油紙壓扁成方形粉片,面上放豆沙或(色條),捲成圓條狀,即可切件進食。

Method

1　Place lard and sugar in plastic tray and melt with boiling water.

2　Add banana essence and mix. Lastly add cooked cake flour (using an egg beater, beat mixture while adding cooked cake flour to make dough).

3　Take 2 taels (80g) of dough and add food colouring. Knead and set aside.

4　Place dough on a large sheet of waxed paper, and press with another sheet of waxed paper into a rectangular thin sheet. Place bean paste or (colour stick) on top and roll into cylinder. Slice and serve.

如意芸豆卷
Ru-yi (Lucky) Sweet Rolls with Kidney Bean Paste

點士 TIPS

白芸豆在國貨公司有售，今次我是從內地購得。此豆營養有益，聞説慈禧太后也十分愛吃芸豆卷。傳説清光緒年間，有一天慈禧太后坐在北海之靜心齋乘涼，忽聽街外有嘈雜聲，命太監出外查究。太監覆奏是芸豆卷和豌豆黃的叫賣聲，慈禧從沒見過，便命人把販商傳入查問。販商下拜，獻上家傳秘製之芸豆卷，慈禧一試極喜，此後販商便被安排進宮中御膳房，專門負責製作芸豆卷和豌豆黃以供慈禧享用，而雲豆卷和豌豆亦成為慈禧最喜愛之點心。

用攪拌器攪豆時，如能攪至極爛，搓至軟滑為最理想。

不一定要包如意形，包成圓形或四方形均可。

Kidney bean is very nutritious and is available in the Chinese department stores. The bean used here was the one I bought from Mainland China. It was said that even Empress Ci-xi loved it very much. The story is that during the Guang-xu era of the Qing Dynasty, when Empress Ci-xi was relaxing at Jingxin Room in Beihai, she heard some noises outside and ordered an eunuch to check it out. The eunuch reported that it was the yellings from a hawker selling kidney bean rolls and peapod cakes. Since she had never seen it before, she sent for the hawker. The hawker bowed before the Empress and offered her the kidney bean rolls made with his family's secret recipe. Ci-xi was surprised and was very pleased with it. After that, the hawker was appointed as a chef in the palace solely responsible for making the two desserts for her, the most favourite dim-sum of the Empress.

Try to make the kidney bean into a mash using a blender.

It is not necessary to make it into a ru-yi shape. You can make some round or square shapes.

材料

- 白芸豆15兩（600克）
- 清水4杯（1000毫升）
- 白糖2.5兩（100克）
- 豬油2湯匙
- 豆沙4兩（160克）

Ingredients

- 600g white kidney bean
- 4 cups (1000ml) water
- 100g caster sugar
- 2 tbsp lard
- 160g red bean paste

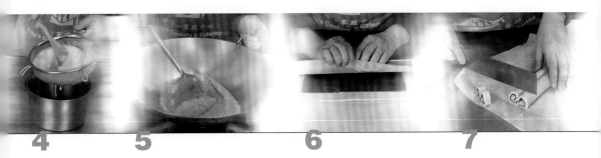

製法

1. 芸豆用水浸4-6小時。
2. 芸豆連水置煲中。煮1小時去衣。
3. 待凍，用4杯清水同置攪拌機中，打爛成豆泥。
4. 壓出水分。
5. 將豬油、白糖、豆泥等放鍋中，炒15分鐘成豆沙後，待凍備用。
6. 將芸豆沙放牛油紙上，壓成長厚片，中心放豆沙，然後兩面捲入成如意卷狀，切件即成。

Method

1. Soak the kidney bean in water for 4 to 6 hours.
2. Put the bean together with the soaking water in a pot. Boil it for an hour. Then remove its skin.
3. When it is cool, put it in a blender. Add 4 cups of water to it and blend it into a bean paste.
4. Press out the water.
5. Put the lard, sugar and bean paste in a wok and saute the mixture for 15 minutes. Leave it to cool and set aside.
6. Put the kidney bean paste on a baking paper and roll it out into a thick sheet. Place the red bean paste in its centre. Roll it up from both ends to make a ru-yi (i.e. lucky charm) shape. Cut it into slices and serve.

材料

糯米粉6兩（240克）

椰汁5兩（200克）

花奶5兩（200克）

白糖4安士（112.5克）

椰茸適量

餡料

豆沙6兩（240克）

Ingredients

6 taels (240g) glutinous rice flour

200g coconut milk

200g evaporated milk

4 oz (112.5g) sugar

Suitable quantities of desiccated coconut

Filling

6 taels (240g) bean paste

製法

1. 將所有材料混合拌勻成粉漿。
2. 把粉漿倒落易潔糕盤中，再用錫紙封口。
3. 置蒸籠內，蒸1/2小時，用手按在粉糰上，不黏手即熟。
4. 戴上透明廚用手套，把蒸好仍熱之粉糰搓至凍（注意：能在溫度高的時候開始搓，效果會最為理想。但須視乎個人手部之耐熱程度，小心燙手）後，取一粉糰包入餡料，再黏上椰茸即成。

Method

1. Mix all ingredients together to form flour paste.
2. Pour flour paste into a non-stick pan and cover with foil.
3. Place in steamer and steam for 1/2 an hour. Test with hand and is cooked when not sticky.
4. With transparent kitchen gloves knead warm dough until it becomes cold. (Note: it is better to knead when the dough is hot. It all depends on the individual as to how hot they can handle. Take care not to burn.) Take pieces of dough and wrap in some filling. Coat with desiccated coconut and serve.

材料A

大菜(剪碎)0.5錢(2克)

山楂餅(壓碎)2兩
(80克)

白糖1兩(40克)

活性乳酸菌飲品
400毫升

清水450毫升

材料B

魚膠粉3湯匙

白糖2湯匙

熱水4湯匙,調溶

Ingredients A

2g agar agar (cut into
small pieces)

2 taels (80g) hawthorn
slices (crushed)

1 taels (40g) white sugar

400ml bio live bacteria
yoghurt drink

450ml water

Ingredients B

3 tbsp gelatine

2 tbsp sugar

4 tbsp hot water, mix
and dissolved

製 法

1　大菜用適量清水浸透(約1小時)盛起。

2　與山楂餅、白糖、清水同置鍋中煮至全部溶解,熄火。

3　拌入已調好之魚膠溶液至均勻無雜質。

4　加入乳酸飲品,拌勻溶合。

5　注入模具中,雪凍凝固即可進食。

Method

1　Soak agar agar in suitable amount of water (about 1 hour) and drain.

2　Boil in pan with hawthorn slices, sugar and water until completely melted.
   Turn off heat.

3　Mix in melted gelatine solution and stir until completely incorporated
   without bits.

4　Mix in yoghurt drink and stir well.

5　Pour into mould and chill until set before serving.

乾蒸蟹黃燒賣
Steamed Crab Roe Siu Mei

TIPS 貼士

由於蟹黃成本較貴，所以食肆選用假蟹黃十分普遍。
此處為各位介紹假蟹黃的做法，以便找不到蟹黃時，
也可以假亂真。

假蟹黃不要煮得太稠，以免硬身難以抹上。

真蟹黃是貴價物料，所以不要抹得太多使其欠真
實感。

餡料打至起膠後置雪櫃略雪後才用，能增加口感。

As real crab roe is very expensive, most restaurants use fake crab roes. The method listed above can be useful when the real stuff is not available.

Do not overcook the fake crab roe for feat that it will become too hard to spread through.

As real crab roes are expensive, do not spread too much on for the authentic look.

Placing the prepared filling in the refrigerator before using to achieve the bite of the final product.

材料
燒賣皮4兩（160克）
蝦仁1/2斤（320克）
瘦肉2.5兩（100克）
豬肉（攪爛）1.5兩
（60克）
肥肉1兩（40克）
冬菇4隻（浸透）

假蟹黃料
蛋黃1隻、橙紅色素少許

調味料
鹽3/4茶匙
糖1/2茶匙
雞粉1/4茶匙
生粉1湯匙
醬油1茶匙
胡椒粉少許
麻油少許
油1茶匙

Ingredients
4 taels (160g) siu mei wrapping

1/2 catty (320g) peeled shrimps

2.5 taels (100g) lean pork

1.5 taels (60g) minced pork

1 tael (40g) pork fat

4 mushrooms (soaked)

Fake Crab Roe Ingredients
1 egg yolk, some yellow red food colouring

Seasonings
3/4 tsp salt

1/2 tsp sugar

1/4 tsp chicken powder

1 tbsp bean flour

1 tsp soy sauce

Dash of pepper

Dash of sesame oil

1 tsp oil

1 **2** **3**
4 **5** **6**

假蟹黃製法
蛋黃放碗中，加入色素攪勻，坐於熱水中，攪
成稠身即成。

製法
1 蝦仁以少許鹽洗擦沖淨，用布吸乾切成細
   粒，瘦肉、肥肉、冬菇均切幼粒。
2 將以上材料置盤中，加入調味撈勻攪至起
   膠，放雪櫃雪約2小時。
3 包成燒賣形，在中央處點上少許假蟹黃，
   放在已塗油之蒸籠或碟中，大火蒸8分鐘
   即成。

Method for Fake Crab Roe
Place egg yolk in bowl and mix with food colouring.
Sit in hot water and stir until thick.

Method
1 Wash peeled shrimps with a little salt. Dry with
   cloth and cut into small cubes. Cut lean pork,
   pork fat and mushrooms into small cubes.
2 Place the above ingredients in bowl and
   add seasonings. Mix until sticky and place in
   refrigerator for about 2 hours.
3 Wrap into siu mei shapes and place a little of
   fake crab roe in the centre. Place in greased
   steamer or plate and steam for 8 minutes over
   high heat.

若把用料中的腐竹改成西洋菜，便成為西菜牛肉。

腐竹要用凍水浸軟。

腐皮又稱鮮竹，將黃豆磨漿，煮熱後把豆漿保持溫度在82℃左右，豆漿表面水分蒸發，大豆蛋白凝結成薄皮狀，將結皮挑起便是濕腐皮。

If use watercress instead of beancurd skin, the dish becomes steamed beef balls with watercress.

Soak beancurd skin in cold water until soft.

Beancurd sheet is made from yellow bean. After boiling the bean milk, keep the temperature around 82°C and part of water will be poured out. Then the bean protein is formed firmly as a tissue layer on the top of milk. Take out the layer by using a pair of long chopstick.

材料	調味料	Ingredients	Seasonings
牛肉1/2斤(320克)	醬油1茶匙	1/2 catty (320g) beef	1 tsp soy sauce
肥肉1兩(40克)	蠔油1湯匙	1 tael (40g) fat pork	1 tbsp oyster sauce
陳皮1片(浸透剁幼)	紹酒1茶匙、雞粉1茶匙	1 pc dried tangerine peel (soaked and minced)	1 tsp Shao Hsing wine, 1 tsp chicken powder
芫茜1湯匙(剁碎)	胡椒粉少許、麻油少許	1 tbsp coriander (minced)	Dash of pepper, dash of sesame oil
馬蹄2粒	生粉1湯匙、清水2湯匙	2 pcs water chestnuts	1 tbsp bean flour, 2 tbsp water
腐皮1張		1 sheet of beancurd skin	

製法

1 牛肉切厚片，先用1茶匙蘇打粉加糖2湯匙撈勻醃透，約40分鐘後沖乾淨(俗稱啤水)，吸乾水分。

2 牛肉再切細粒，與肥肉及馬蹄一起剁爛，加入陳皮、芫茜及調味等攪勻至彈性。

3 撻透放雪櫃，雪凍半天才用最好。

4 腐皮抹淨剪開炸透，以水略浸後，用毛巾吸乾置碟中。

5 將牛肉取出捏成小球形，放在腐皮上，以大火蒸5分鐘即成。

Method

1 Slice beef in thick pieces and mix with 1 teaspoon of bicarbonate of soda and 2 tablespoons of sugar. Marinate for about 40 minutes and rinse (commonly known as water rinsing). Absorb all moisture.

2 Cut beef into small cubes and mince with fat pork and water chestnuts. Add dried tangerine peel, coriander and seasonings. Stir and mix until elastic.

3 Slap mixture thoroughly and place in refrigerator. Chill for at least half a day before using.

4 Wipe beancurd skin, cut into pieces and deep fry. Soak in water and dry with towel. Place on plate.

5 Remove beef and form into small balls. Place on top of beancurd skins and steam on high heat for 5 minutes.

貼士 TIPS

這是一款在普通紮蹄中心加入蝦子的懷舊小吃。

買回來的蝦子，必須用白鑊炒熟才用，否則會有腥味；炒好後可放入罐或瓶中，需要時取出適合用量即可。

包製紮蹄，可用牛油紙、布或麵粉袋等。用布紮作，形狀較佳，但如不常用，於起出製成品時，紮布容易把紮蹄外形弄破。牛油紙沒有此毛病，容易用，但需要常加練習，才可紮到美觀外形。

This is a traditional snack with the addition of shrimp roe in a normal version of tied rolls.

The bought shrimp roe must be fried in a dry pan before using. Otherwise they will smell fishy. After frying they can be stored in a tin or bottle and use when required.

When tying rolls, wax paper, cloth or flour sacks can be used. Using cloth gives a better shape but if the cloth is not used regularly, it may damage the outside of the roll. Wax paper does not have this problem. It is easy to use but must be practised often to make a perfect looking tied roll.

材料	調味料	Ingredients	Seasonings
鮮竹4塊、薑4片	醬油(或老抽)2湯匙	4 sheets fresh bean curd skins, 4 slices ginger	2 tbsp soy sauce (or dark soy sauce)
白繩2條	蠔油1湯匙	2 pcs white string	1 tbsp oyster sauce
扣布或牛油紙	片糖1/2磚	Muslin or waxed paper	1/2 block slab sugar
蝦子(白鑊炒香)1兩	雞粉1湯匙	1 tael (40g) shrimp roe (fried in dry pan)	1 tbsp chicken powder
(40克)	油3湯匙		3 tbsp oil
	清水2杯(500毫升)		2 cups (500ml) water

製法

1. 腐皮雙摺成半圓形，剪為2份(共8塊)。
2. 調味料煮滾收慢火，先將4塊腐皮拖軟盛起。
3. 將其餘4塊加入調味料中，慢火煮至乾身。
4. 將拖軟之腐皮打開，覆上已煮乾之腐皮，中心放蝦子，從尖角開始捲成長卷，以繩紮成紮蹄，隔水蒸1/2小時即成。

Method

1. Fold bean curd skins into halves to form semi-circles. Cut into 2 (total 8 pieces).
2. Boil seasonings and lower heat. Put in 4 sheets of bean curd skin until soften. Remove.
3. Place the remaining 4 sheets into seasonings and cook on low heat until dry.
4. Open softened bean curd skins and cover with dried bean curd skins. Place shrimp roe in centre and rolling from the tip, make into a cylinder shape. Tie with string into a tied roll. Steam over water for 1/2 an hour.

薑味朱古力湯丸（花好月圓）

Ginger Flavoured Chocolate Dumplings Flowers Blooming Under a Full Moon

餡料

- 純朱古力4兩(160克)
- 棉花糖1 3/4兩(70克)
- 無鹽花生1.5兩(60克)

皮材料

- 糯米粉1/2斤(320克)
- 粘米粉4湯匙
- 清水12安士(340毫升)

薑味糖水

- 薑1兩(40克)
- 清水2杯(500毫升)
- 黃糖1湯匙

Filling

- 4 taels (160g) plain chocolate
- 70g marshmallows
- 60g peanuts with no salt

Wrapping

- 1/2 catty (320g) glutinous rice flour
- 4 tbsp rice flour
- 12 oz (340ml) water

Ginger Flavoured Syrup

- 1 tael (40g) ginger
- 2 cups (500ml) water
- 1 tbsp brown sugar

貼士 TIPS

由於朱古力味甜，故糖分略少而薑汁較重(為取薑之香氣)。

以隔水加熱的方式坐溶朱古力，水的溫度最合適是50-60℃，如超過此溫朱古力便很難溶解。

注意要避免隔水加熱時，所用的溫水滲入朱古力中。

As chocolate is quite sweet, it is better to put more ginger juice, which enhances the gragrant taste, than sugar.

The best way to melt chocolate is the double boiling method. The water temperature should be kept at 50-60°C, if over this degree, the chocolate cannot be melted.

Note to avoid the mix of chocolate with the water.

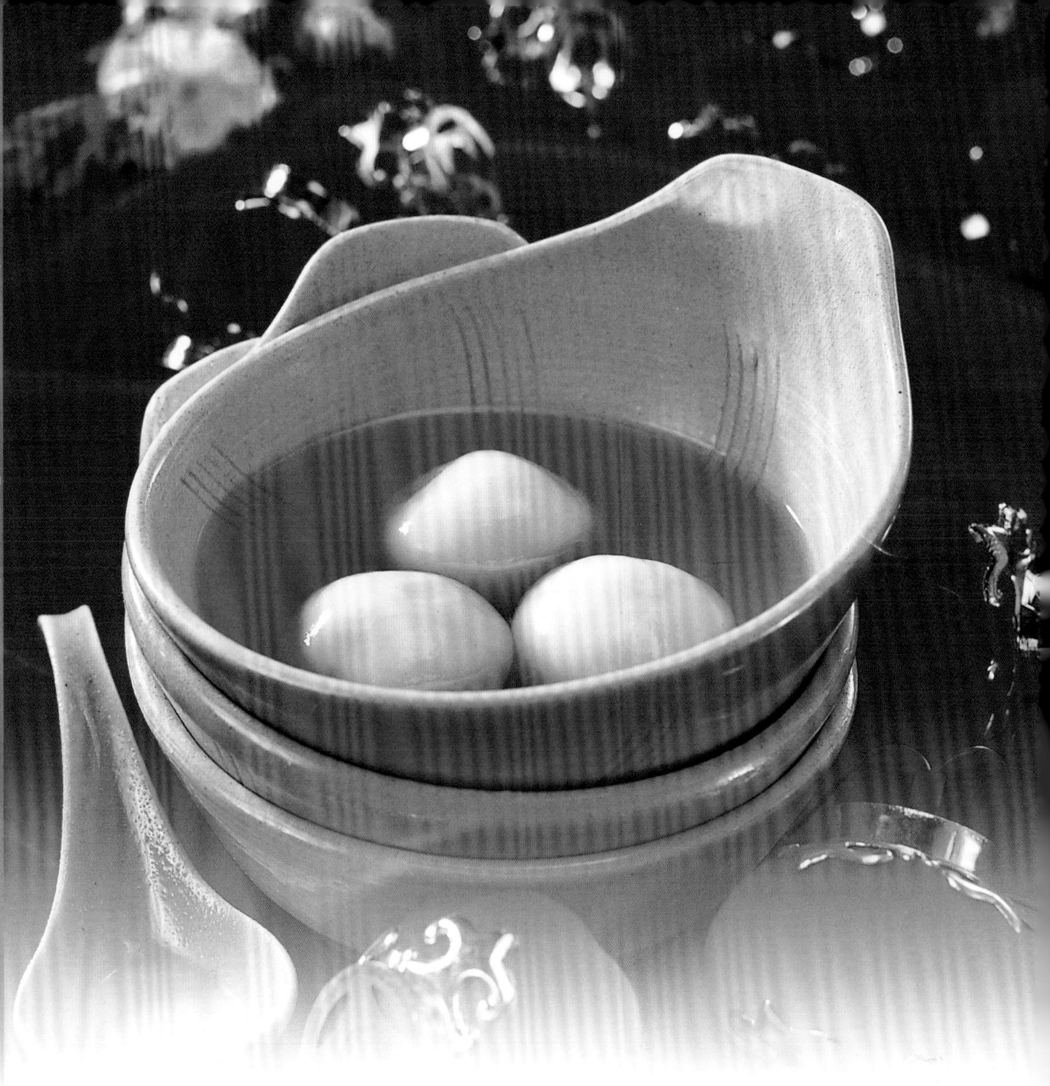

餡料製法 Method for Filling　　　　　　　　　薑味糖水製法 Method for Ginger Flavoured Syrup

製法 Method

餡料製法

1. 純朱古力切碎，坐於溫水中至溶解。
2. 將花生碎粒、棉花糖加入朱古力溶液中，混和後倒落四方盆中，雪凍，切成小粒搓圓備用。

薑味糖水製法

1. 薑去皮磨茸，榨汁。
2. 將薑汁與黃糖及水煮沸備用。

製法

1. 糯米粉、粘米粉篩入盆中，將清水慢慢加入搓成軟硬適中之粉糰，分成30小粒。
2. 將粉糰捏一深窩，放入1粒棉花糖朱古力收口搓圓。
3. 燒半鍋水，水滾投入糯米球，大滾再加凍水1杯，待再滾起見浮者便可撈出，置碗中，加入薑味糖水進食。

Method for Filling

1. Chop chocolate and place on top of warm water to melt.
2. Chop peanuts and place in melted chocolate with marshmallows. Mix well and place in a square pan. Chill and cut into pieces. Roll pieces into balls and set aside.

Method for Ginger Flavoured Syrup

1. Peel ginger and grate. Squeeze out juice.
2. Boil ginger juice with brown sugar and water. Set aside.

Method

1. Sieve glutinous rice flour and rice flour into bowl. Add water gradually and stir until soft dough is formed. Divide into 30 small balls.
2. Make a deep dent in each ball and place in a piece of chocolate and marshmallow. Close gap and roll into a ball.
3. Heat half a pan of water and when it boils add glutinous rice flour balls. Add a cup of cold water when water comes to a rapid boil. When water comes to a boil again, remove dumplings that have floated to the top. Place in bowl with ginger syrup and serve.

上海芝麻湯丸
Shanghainese Sesame Seed Dumplings

最好選用能放入全部湯丸的湯鍋為佳。

將湯丸放入沸水中，不應攪動，以免使其爆裂及黏底。

當水再滾起，此時縱然有部分湯丸浮面，也並不表示熟透，應加入1杯約8安士之凍水，及至再度水沸時，浮上的湯丸便熟得恰到好處。

Choose a pan big enough to accommodate all the dumplings.

Do not stir dumplings after they have been placed in the boiling water. This may cause them to split or stick to the bottom.

When the water comes to the boil, some dumplings may float to the top. This does not mean that they are cooked. A cup of cold water, about 8 oz, should still be added. When the water comes to the boil at the second time, the dumplings are then completed cooked.

材料	餡料	Ingredients	Filling
糯米粉 1/2 斤（320 克）	芝麻糖粉（黑羊酥）	1/2 catty (320g) glutinous rice flour	1.5 taels (60g) sesame icing sugar
粘米粉 4 湯匙	1.5 兩（60 克）	4 tbsp rice flour	1.5 taels (60g) white sugar
水 12 安士（340 毫升）	砂糖 1.5 兩（60 克）	12 oz (340ml) water	2 taels (80g) pork fat
	豬膏（板油）2 兩（80 克）		

製法

1. 豬膏（板油）洗淨，用刀剁至極爛。
2. 加入砂糖和黑羊酥拌勻搓透做成小圓粒，放雪櫃冷凍至硬實備用。
3. 糯米粉、粘米粉篩入盆中，以適量清水搓成軟硬適中之粉糰，再搓成長條狀，分切成 30 小粒。
4. 將粉糰捏一深窩，放入一粒餡料，收口搓圓。
5. 燒半鍋水，水滾投入糯米球，大滾再加 1 杯水，待再滾起見浮者便可撈出。

Method

1. Wash pork fat thoroughly and mince with knife.
2. Add white sugar and sesame icing sugar and mix well. Form into small balls and place in refrigerator.
3. Sieve flour and glutinous rice flour into bowl. Add sufficient water to form a soft dough. Form into roll and cut into 30 small pieces.
4. Make a dent in each piece and place in a ball of filling. Wrap dough round filling to enclose. Roll into a round dumpling.
5. Heat half a pan of water and when boiling place in dumplings. When water comes back to the boil, add a cup of cold water. Bring water back to the boil. Remove dumplings when they float to the top.

舊日的蝦多士，材料中包含豬肉及薯茸。並非如現今流行的，以全蝦肉製成。喜愛海鮮的人，可能有點兒失望，但偶爾品嚐另類製作去懷舊一番，亦一樂也。

要注意的是，油必須滾；餡料不要塗得太厚，以免炸時難熟透；先炸餡料的一面至金黃色，反另一面再炸至金黃即可。

The traditional shrimp toast contains pork and mashed potato. The modern version has pure shrimp meat. This may not appeal to seafood lovers but having a taste of traditional food now and then is one of life pleasures.

Note that the oil must be hot and filling not too thick. Otherwise it is difficult to cook through. Fry the filling side first until golden brown before turning over to the other side.

材料

蝦仁4兩（160克）
豬肉（攪爛）2兩（80克）
中蝦（去殼留尾起雙飛）
32隻
洋蔥粒2湯匙
薯仔（煮腍搓爛）1個
雞蛋2隻、麵包糠適量
麵包1/2磅

調味料

鹽1/2茶匙
雞粉1茶匙
糖1/2茶匙
生粉2茶匙
胡椒粉少許

Ingredients

4 taels (160g) peeled shrimps

2 taels (80g) pork (minced)

32 pcs medium size shrimps (peel with tails on and split)

2 tbsp chopped onion

1 potato (cooked and mashed)

2 eggs, Breadcrumbs

1/2 lb bread

Seasonings

1/2 tsp salt

1 tsp chicken powder

1/2 tsp sugar

2 tsp bean flour

dash of pepper

製法

1　蝦仁洗淨，吸乾水分剁爛成茸。

2　將蝦膠、豬肉加入調味撈勻攪透，再加入洋蔥及薯茸、雞蛋1隻攪勻。

3　麵包一開四，再切三角形共8件，另1隻雞蛋打散。

4　將適量餡料釀在麵包上，塗上少許蛋液，面上釀上1隻鮮蝦，再上麵包糠。

5　將做好之麵包放中火油內，炸成金黃色，瀝油即可進食。

Method

1　Wash peeled shrimps, dry thoroughly and mash.

2　Mix mashed shrimps with pork and seasonings. Mix well and add onion, mashed potato and a beaten egg.

3　Cut bread into quarters and then triangles making 8 pieces in total. Beat the remaining egg.

4　Place suitable amount of filling on bread and brush with beanten egg. Place a shrimp on top and coat with breadcrumbs.

5　Place prepared shrimp toasts in medium hot oil and deep fry until golden brown. Drain and serve.

苔條花生
Seaweed Peanuts

材料
花生仁1/2斤（320克）
苔條1兩（40克）
味椒鹽適量

調味料
清水2杯（500毫升）
糖4湯匙

Ingredients
1/2 catty (320g) shelled peanuts
1 tael (40g) seaweed
Chilli salt to taste

Seasonings
2 cups (500ml) water
4 tbsp sugar

製 法
1 花生仁洗淨放鍋中，加入調味拌勻，以中火煮3分鐘至汁水收乾。
2 盛起瀝乾水分，放於當風處吹乾。
3 苔菜用清水洗淨，用布吸乾水分。
4 用大火燒至高溫，倒下苔菜炸20秒即盛起壓碎備用。
5 燒油至微溫，放入花生炸至浮起呈金黃色盛起瀝油。
6 將苔菜撒上，再加椒鹽拌勻即成。

Method
1 Wash peanuts and place in pan. Add seasonings and mix well. Cook on medium heat for 3 minutes until the liquid has been absorbed.
2 Remove and drain. Place in draft and allow to dry.
3 Wash seaweed and dry with cloth.
4 Using high heat until very hot, deep fry seaweed for 20 seconds. Drain and crush.
5 Heat oil to medium hot and add peanuts. Deep fry until they float and are golden brown. Remove and drain.
6 Scatter on seaweed and mix with chilli salt.

材料
腰果1斤(640克)
味椒鹽少許

調味料
海鮮素1湯匙
咖喱粉2湯匙
糖4湯匙
麥芽糖4湯匙
清水2杯(500毫升)

Ingredients
1 catty (640g) cashew nuts
A little spiced salt

Seasonings
1 tbsp seafood flavouring
2 tbsp curry powder
4 tbsp Sugar
4 tbsp Malt
2 cups (500ml) water

製 法
1 腰果洗淨，加調味拌勻置鍋中，以中火煮滾約3分鐘，至湯汁收乾。
2 盛起置當風處，吹至極乾備用。
3 燒熱鑊放油，燒至微溫，將腰果傾下。
4 以慢火炸至腰果浮起呈金黃色，盛起待涼。
5 撒上味椒鹽即成。

Method
1 Wash cashew nuts, mix with seasonings and place in pan. Boil on medium heat for 3 minutes until the liquid is absorbed.
2 Dish and place in draughty area until completely dry.
3 Heat wok and add oil. When oil is slightly warm add cashew nuts.
4 Fry on low heat and stir until cashew nuts float to the top. Drain and allow to cool.
5 Sprinkle on spiced salt and serve.

材料

花生肉1斤(640克)

南乳1塊

調味料

鹽2茶匙

糖1湯匙

八角2粒

水2杯(500毫升)

Ingredients

I catty (640g) shelled peanuts

I pc red preserved bean curd

Seasonings

2 tsp salt

I tbsp sugar

2 star anise

2 cups (500ml) water

製法

1　將調味料開勻。

2　花生洗淨放入南乳及調味料中，浸約1/2小時(須不時加以攪動)，使其入味。

3　將浸好之花生盛起瀝乾，放當風處吹約3-4小時至花生外衣極乾爽。

4　將花生放入鑊中以慢火炒熟，或慢火在焗爐烘至脆亦可(約15分鐘)。

Method

1　Mix seasonings ingredients.

2　Wash peanuts and place in red preserved taro curd and seasonings solution. Soak for 1/2 an hour (stirring from time to time) to absorb the flavour.

3　Drain soaked peanuts and place in draughty area for about 3-4 hours until the outside of the peanuts are very dry.

4　Place peanuts in wok and stir fry over slow heat until cooked. Alternately, Place in a slow oven and bake until crispy (about 15 minutes).

材料

糯荔甫芋1斤（640克），
切幼條

糯米粉 1/4 杯

粘米粉 1/4 杯

泡打粉 1/3 茶匙

調味料

酒 1 湯匙

鹽 1/2 茶匙

五香粉 1/2 茶匙

Ingredients

1 catty (640g) laipo taro,
shredded

1/4 cup glutinous rice
flour

1/4 cup rice flour

1/3 tsp baking powder

Seasonings

1 tbsp wine

1/2 tsp salt

1/2 tsp five spice powder

芋蝦
Crispy Taro Balls

製 法

1 芋頭去皮洗淨，用乾布抹至極乾後，薄切幼條。

2 將調味加入芋條中，輕輕拌勻片刻，待軟身後，再加入已篩之粉料撈勻。

3 燒熱油，放下罩籬炸熱，把適量之芋條放在罩籬中，不可壓實，放八成滾油中以中火炸脆，炸時要篩動罩籬，炸至硬身，芋蝦便會自動脫出模型。

Method

1 Peel and wash taro. Wipe with dry cloth until completely dry. Shred finely.

2 Add seasonings to shredded taro and mix lightly. When soften, add sieved flour and mix well.

3 Heat oil and place in wire ladle. Place a suitable amount of shredded taro in ladle. Do not press hard. Place in 80% hot oil and deep fry on medium heat until crispy. Move ladle around when frying and when the taro is cooked, it will fall out of the ladle automatically.

貼士 TIPS

蠔仔煎即潮式蠔餅。因蠔表面有潺，清洗時，必須以鹽、生粉及熟油揉擦才會清潔及有香氣。

飛水為保險做法。

鴨蛋與蕃薯粉的比例為1:10，鴨蛋有質感及黏性，配以蕃薯粉，口感特佳。

油量要多才會香脆。

This is the same as the Chiu Chow pan fried oysters. As there is slime on the oysters, they must be washed with salt, bean flour and cooked oil. Rub gently for cleaning and spreading the aroma of oysters.

Blanching in water is a safety procedure.

The ratio between duck egg and potato water is 1:10. Duck egg has more texture and stickiness. Team with sweet potato flour to bring fabulous chew more bite.

Use plenty of oil to achieve crispiness.

材料	調味料	Ingredients	Seasonings
蠔仔4兩(160克)	鹽1/2茶匙	4 taels (160g) baby oysters	1/2 tsp salt
鴨蛋1隻(打散)	雞粉1茶匙	1 duck egg (beaten)	1 tsp chicken powder
葱2條(切粒)	麻油少許	2 pcs spring onion (diced)	Dash of sesame oil
香芹1條(切粒)	胡椒粉少許	1 stick Chinese celery (diced)	dash of pepper

薯粉水料	Potato water solution
薯粉10湯匙(115克) 清水8湯匙(92毫升)，開勻	10 tbsp (115g) potato flour mix with 8 tbsp (92ml) water

1 2 3

4 5

製法

1 蠔仔用生粉及生油洗淨。
2 飛水瀝乾備用。
3 將調味料與薯粉水開勻加入蠔仔及葱粒。
4 鴨蛋打起倒落蠔仔中拌勻。
5 燒熱鑊下油3湯匙，下一杓蠔仔料，搪勻煎成兩面金黃色即成。

Method

1 Wash baby oysters with bean flour and oil.
2 Blanch in boiling water. Drain and set aside.
3 Mix seasonings with potato water. Add to baby oysters and spring onion.
4 Beat duck egg and add to baby oyster mixture. Mix well.
5 Heat wok and place in 3 tablespoons of oil. Pour in one ladle of baby oyster mixture. Swirl and fry until both sides are golden brown.

炸春卷要有耐性，由於包入餡料時會捲上幾層，所以表面看似金黃，但內層可能仍有濕氣而未能炸透，凍後便會回軟。故應多炸一點時間才取出(炸約10分鐘也無妨)；這樣，凍後不但鬆脆可口，若以保鮮袋保存入罐，更可存放2-3天之久。

Patience is required when frying spring rolls. As there are several layers of wrapping rolled up, the outside may look golden brown while the inside may still be moist. This will make the spring roll soft after cooling. Fry a little longer before removing (fry up to about 10 minutes). After cooling, spring roll will be crispy and can be kept for 2-3 days in ziplock bags with a tin outside.

點士 TIPS

材料	調味料	Ingredients	Seasonings
芋頭1/2 斤（320克）	醬油1湯匙	1/2 catty (320g) taro	1 tbsp soy sauce
濕發冬菇1兩（40克）	蠔油1湯匙	1 tael (40g) soaked mushrooms	1 tbsp oyster sauce
沙葛1/2 斤（320克）	鹽1/2茶匙	1/2 catty (320) jicama	1/2 tsp salt
蝦米1兩（40克）	糖1茶匙	1 tael (40g) dried shrimps	1 tsp sugar
蒜茸1茶匙	雞粉1茶匙	1 tsp crushed garlic	1 tsp chicken powder
春卷皮1斤（640克）	麻油少許	1 catty (640g) spring roll wrapping	dash of sesame oil
五香粉1茶匙	胡椒粉少許	1 tsp five spice powder	dash of pepper

1

2

3

4

製法

1. 將以上所有材料洗淨，浸透，切絲。
2. 芋頭切絲後，先用八成熱油，炸至金黃盛起。
3. 燒熱2湯匙油，炒香蒜茸，傾下冬菇、蝦米炒香，加入芋絲及沙葛、五香粉等炒透，最後加入調味兜勻盛起待凍備用。
4. 春卷皮攤開，包入適量餡料向外捲起，以粉漿糊口。
5. 燒油約八成熱，將春卷放下以中慢火，炸至金黃內外全脆（約10分鐘）即成。

Method

1. Wash all the above ingredients, soak and shred.
2. Deep-fry theshredded taro in 80% hot oil until golden brown.
3. Heat 2 tablespoons of oil and fry crushed garlic. Add mushrooms and dried shrimps. Then put in shredded taro, jicama and five spice powder. Mix well and lastly put in seasonings. Stir fry until mixed. Dish and cool.
4. Spread spring roll wrapping and place some filling on top. Roll outwards and seal with flour solution.
5. Heat oil to 80% hot. Place spring rolls into oil. Fry on medium low heat until golden brown and crispy both on the inside and out (about 10 minutes).

炸蘿蔔絲餅又稱蘿蔔油池，是香港流行之街頭小吃之一，其麵漿可用作炸其他食品(如炸蕃薯、炸芋頭等)之脆漿。調好之麵漿不應立即使用，需靜置半小時待其發酵才會脆。

油提模必須先行放滾油中燒熱，才能注入麵漿，否則難起模。材料八分滿即可，太滿容易溢出。

蘿蔔刨絲或切絲均可。拌勻調味後會滲出水分，因此應將水分盡量壓出，才不至影響麵漿之濃度。

炸製需時10分鐘以上，以免材料不熟。油溫不可太高，否則表面燒焦而內裏熟不透，中慢火即可。炸至定形後，餅會自動脫模。

Fried shredded turnip cake is also known as oil pool turnip. It is one of the favourite street snacks. The flour mixture can be used as a batter for other ingredients (e.g. sweet potato, taro etc.). The prepared flour mixture should not be used immediately. Leave for 1/2 an hour for it to rise to give a crispy texture.

The mould for frying must be heated in the oil before adding flour mixture. Otherwise the finished cake will not drop out. Fill no more than 80% to prevent overflowing.

Turnip can be shredded or cut into shreds. After mixing with the seasonings, water will be excreted. Squeeze all water out as not to affect the density of the mixture.

Frying time is over 10 minutes to prevent undercooking. Temperature must not be too hot otherwise the surface will be too brown before the inside is cooked. Fry on medium low heat. When the cake is set, it will drop out of the mould automatically.

材料
麵粉 1/2 斤（320 克）
粟粉 2 兩（80 克）
發粉 2 茶匙
清水 2 杯（500 毫升）
雞蛋 1 隻
油 1/4 杯

餡料
蘿蔔 1 斤（640 克）
蝦米 1 兩（40 克）剁碎
葱 2 條（切粒）

調味料
鹽 1 茶匙
雞粉 1 茶匙
胡椒粉少許
糖 1 茶匙

Ingredients
1/2 catty (320g) flour
2 taels (80g) corn flour
2 tsp baking powder
2 cups (500ml) water
1 egg
1/4 cup oil

Filling
1 catty (640g) turnip
1 tael (40g) dried shrimps,
minced
2 pcs spring onion (diced)

Seasonings
1 tsp Salt
1 tsp chicken powder
dash of pepper
1 tsp sugar

1 2 3 4

5 6 7 8

製法
1　蘿蔔去皮洗淨刨絲。
2　加入調味拌勻，醃至軟身備用。
3　麵粉、粟粉及發粉同篩勻。
4　雞蛋打起，將麵粉混合料與清水分 3 次加入，每加 1 次必需打勻至加完為止。
5　最後加油攪勻成麵糊。
6　將醃好之蘿蔔和蝦米、葱等加入麵糊中攪透備用。
7　燒滾油，將長柄油提放入浸熱。取出加入蘿蔔絲粉漿約八成滿。
8　放回油鍋內炸片刻，待粉漿定型後自動退出油提外，繼續炸至金黃（約 10 分鐘）即成。

Method
1　Peel turnip, wash and shred.
2　Add seasonings and mix well. Marinade until soften.
3　Sieve flour, corn flour and baking powder.
4　Whisk egg until risen. Mix in sieved flour in three batches. Whisk until completely incorporated after each addition.
5　Lastly put in oil and mix to flour paste.
6　Add marinated turnip, dried shrimps and spring onion etc. and mix well.
7　Heat oil till hot, place in long handle ladle to get hot. Remove and fill up to 80% of turnip mixture.
8　Return to oil and fry briefly. When the flour mixture is set, it will drop out of ladle. Continue frying until golden brown (anout 10 minutes). Serve.

酸梅湯
Sour Plum Drink

材料

酸梅湯料1包
冰糖1斤 (640克)
清水8杯 (2000毫升)

Ingredients

1 packet sour plum drink ingredients
1 catty (640g) rock sugar
8 cups (2000ml) water

製 法

1　將湯料略洗放煲中。
2　注入清水，沸後以慢火熬至剩餘4杯。
3　將冰糖加入，再熬剩2杯時。
4　待凍隔去渣盛起，放雪櫃。
5　如飲用時可將一半酸梅湯加一半凍滾水開均便可飲用。

Method

1　Wash drink ingredients briefly and place in pan.
2　Pour in water and when boiling, turn heat to low and simmer until 4 cups of liquid are left.
3　Add rock sugar and simmer until 2 cups of liquid are left.
4　When cold, drain and preserve liquid. Place in refrigerator.
5　When serving, dilute drink with equal quantity of cold drinking water.

材料

西米1兩(40克)

圓肉1/2兩(20克)

糖冬瓜1兩(40克)切粒

百果1.5兩(60克)去殼及衣計

冰糖適量

蓮子1.5兩(60克)

紅棗1兩(40克)

柿餅1.5兩(1個60克)切薄片

鵪鶉蛋適量

Ingredients

I tael (40g) sago

1/2 tael (20g) dried logan meat

I tael (40g) sugared winter melon

1.5 taels (60g) gingko nuts, shelled and skin removed rock

Sugar to taste

1.5 taels (60g) lotus seeds

I tael (40g) red dates

1.5 taels (1 weighing 60g) persimmon sliced thinly

Some quail eggs

製法

1 西米用清水浸1/2小時,瀝乾放水中煮至透明,過冷備用。

2 蓮子浸透後用清水煮10分鐘去芯備用。

3 圓肉、紅棗沖淨飛水備用。

4 鵪鶉蛋預先焓熟,浸凍水後去殼備用;柿餅切薄片。

5 將以上各材料(除鵪鶉蛋、蓮子、西米外)同置鍋中,注入適量清水煮1/2小時,加入冰糖。

6 再把餘下之材料傾下,煮至糖溶即可進食。

Method

1 Soak sago for 1/2 an hour. Drain and cook in water until transparent. Rinse in cold water and set aside.

2 Soak lotus seeds thoroughly and boil in water for 10 minutes. Remove hearts and set aside.

3 Wash logan fruit and red dates, blanch in water and set aside.

4 Boil quail eggs until cooked. Soak in cold water and peel. Set aside. Thinly slice dried persimmon .

5 Place the above ingredients (except quail eggs, lotus seeds and sago) in pan. Add sufficient water and boil for 1/2 an hour. Add rock sugar.

6 Place in remaining ingredients and boil until sugar melts. Serve.

烏梅楊桃蜜

Black Plum & Star Fruit Drink

材料

楊桃4個
烏梅4粒
砂糖 1/2 杯（或適量）
桂花糖1湯匙
梅酒 1/4 杯
清水2杯（500毫升）

Ingredients

4 pcs star fruit

4 pcs black plums

1/2 cup sugar (or to taste)

1 tbsp osmanehus sugar

1/4 cup plum wine

2 cups (500ml) water

製 法

1　楊桃洗淨切片。
2　烏梅用清水浸約2小時。
3　加糖煮剩1.5杯至烏梅出味待凍。
4　切了星片狀之楊桃浸漬在糖水中。
5　加入桂花糖和梅酒，雪凍即可飲用。

Method

1　Wash star fruit and slice.
2　Wash black plums and soak in water for 2 hours.
3　Add water and boil until 1.5 cups of water is left and the black plum flavour is released. Allow to cool.
4　Soak sliced star fruit in sugar solution.
5　Add osmanehus sugar and plum wine. Chill and serve.

材料

涼粉草 113 克
食用鹼水 1 3/4 湯匙
清水 18 杯 (4.5 公升)

茨料

鷹粟粉 (或生粉) 125 克
清水 2.5 杯 (625 毫升)

Ingredients

113g mesona chinensis
1 3/4 tbsp alkaline water
18 cups (4.5L) water

Gravy

125g cornflour
2.5 cups (625ml) water

製法

1. 涼粉草洗淨置煲中，注入 18 杯清水，加入鹼水 1 3/4 湯匙，煲滾後，改以中慢火煲 2 小時。

2. 涼粉水略凍，用力把涼粉草擦出膠質，再煲 1/2 小時，又再重複擦一次，然後用篩隔去渣，再以布袋隔一次 (此時應剩回涼粉水 12 杯)。

3. 將茨料開勻，倒進滾起之涼粉水中，迅速攪勻，然後倒入一個已用冰水搪過之盆中，冷後雪凍，便可切粒加入糖水食用。

Method

1. Wash the mesona chinensis thoroughly and then put it in a pot. Pour in 18 cups of water, add in 1 3/4 tbsp of alkaline water and bring it to a boil. Then simmer it under medium heat for 2 hours.

2. When the mesona chinensis liquid is cool, rub the mesona chinensis vigorously to remove its gel. Then boil it for another 1/2 hour. Repeat the rubbing and boiling processes. After that, filter out the residue with a sift and then filter it again with a porous cloth bag (leave only about 12 cups of mesona chinensis liquid behind).

3. Mix the gravy ingredients, pour it into the grass jelly solution and stir it quickly. Then pour it into a mixing bowl rinsed with iced water. Chill it in a refrigerator. Dice and mix it with syrup before serving.

杏仁豆腐

大菜 1/4 安士（7.5克）

鮮豆漿 2.5杯（625毫升）

糖 1/2杯

杏仁香油 1/2茶匙

雜果 1/2杯

清水 2杯（500毫升）

Ingredients

1/4 oz (7.5g) agar agar

2.5 cups (625ml) fresh soya milk

1/2 cup sugar

1/2 tsp almond essence

1/2 cup mixed fruit

2 cups (500ml) water

製 法

1 大菜洗淨，用清水浸軟。
2 鍋中注入2杯清水，加入瀝乾水分之大菜，以中慢火煮至溶解。
3 加糖煮溶後熄火注入鮮豆漿和香油攪勻。
4 用篩隔去雜質，倒入盆中待凍置雪櫃。
5 凝固後便可取出切粒或舀於碗中拌入雜果進食。

Method

1 Wash agar agar and soak in water until soft.
2 Place 2 cups of water in pan and add drained agar agar. Cook on medium low heat until it dissolves.
3 Add sugar and boil until melted. Turn off heat and place in fresh soya milk and essence. Mix well.
4 Sieve away any impurities and pour into pan. Place in refrigerator when cold.
5 When set, remove and cut into cubes. Add mixed fruit and serve.

材料

冰糖 1 3/4 兩（70克）

清水 1 杯（250毫升）

雞蛋 3 隻

綠茶粉 2 湯匙

Ingredients

1 3/4 taels (70g) rock sugar

1 cup (250ml) water

3 eggs

2 tbsp green tea powder

製法

1　將冰糖放入清水中煮溶，加入綠茶粉。

2　雞蛋打起，將熱糖水撞落蛋液內，邊撞邊打勻。

3　以篩隔去雜質後倒入小碗中。

4　置蒸籠以慢火及八分蓋燉10分鐘即成。

Method

1　Boil rock sugar in water until melted. Add green tea powder.

2　Whisk eggs and pour hot sugar solution into eggs, whisking all the time.

3　Sieve to remove all impurities and place in small bowl.

4　Place in steamer and steam on low heat with cover 20% ajar for 10 minutes.

醃酸薑
Pickled Ginger

材料
子薑 1斤（640克）

糖醋料
洋醋 2杯（500毫升）
糖 1.5杯
鹽 1/2 茶匙

Ingredients
1 catty (640g) young ginger

Sugar & Vinegar Sauce Ingredients
2 cups (500ml) distrilled vinegar
1.5 cups sugar
1/2 tsp salt

製法
1. 子薑去皮切薄片（或切角均可）。
2. 用少許鹽（約1茶匙）拌勻，30分鐘後清洗乾淨，瀝乾水分。
3. 糖、醋置煲中，慢火煮至糖溶待凍。
4. 將子薑放落糖醋中，醃一天便可食用。

Method
1. Peel young ginger and cut into thin slices (or triangles).
2. Mix with a little salt (about 1 teaspoon). Wash after 30 minutes and drain.
3. Place sugar and vinegar sauce ingredients in pan and simmer on low heat until sugar melts. Allow to cool.
4. Add young ginger to sauce and marinate for 1 day. Serve.

材料
雞蛋36隻
花紅粉3茶匙及
熱水1杯，開溶備用

Ingredients
36 eggs

3 tsp red flower food
colouring & 1 cup hot
water (mixed & set aside)

製法
1 花紅粉放熱水中。
2 將雞蛋放凍水中，以中火煮滾。
3 約20分鐘取出。
4 趁熱放入花紅粉水中上色。

Method
1 Place red flower food colouring in hot water.
2 Place eggs in cold water and boil on medium heat.
3 Remove after about 20 minutes.
4 Place in red flower solution while hot for easy colouring.

豬腳薑
Pickled Pork Knuckles with Ginger

點 | TIPS

在甜醋還未面世的時候，前輩的長者們，都是以酸醋加糖且一般不加水分的方法去製作薑醋。其後添丁甜醋的出現，因十分方便，古老的方法已漸漸被人遺忘。

用甜醋雖然可省去很多麻煩和時間，但製成品卻不及老方法可口，由於某些品牌的甜醋，其甜味多來自糖精，也不利於健康。此處除介紹原始的舊方法外，更加入了我在星洲居住時當地人的製作心得，便是在醋中加入水分。加水的好處，能令薑肉鬆化、汁液不至太濃而產生膩口感覺。相反，若不加水，薑的纖維會因久煮而收緊，質地變得韌而實，欠鬆化。大家不妨試作比較。

豬腳先用清水煮，使其去除多餘油分及易腍。

太早加入糖分，會使物料不易腍。因此先將物料煮腍，翌日才下糖是理想做法。

Before the introduction of sweet vinegar, the older generation usually add sugar but no water to normal vinegar to make ginger vinegar. With the appearance of sweet vinegar for new babies, the old method is gradually forgotten as it is so convenient.

Although using sweet vinegar saves a lot of time but the taste is not as good as the traditional ones. Some brands of sweet vinegar use saccharin which is not too good for health. The method introduced here is the original traditional method. It also includes a few tips I picked up from the locals when I was living in Singapore. They add water to the vinegar. The advantage is to fluff up the texture of the ginger and the liquid sauce is not too thick to be filling. On the other hand, if water is not added, the fibres in the ginger will contract during cooking making it tight and tough. It is advisable to try and compare.

Boil pork knuckles in water first will rid them of excess fat and easier to become tender.

If sugar is added too early, the ingredients will be hard to become tender. So it is better to cook the ingredients until tender and add sugar the next day.

材料

黑酸醋2斤（1280克）	
清水1斤（640克）	
片糖3斤（1920克）	
薑3斤（1920克）	
豬腳（斬件）2隻	
雞蛋10隻	

Ingredients

- 2 catties (1280g) black vinegar
- 1 catty (640g) water
- 3 catties (1920g) sugar slabs
- 3 catties (1920g) ginger
- 2 pork kunckles (cut into pieces)
- 10 eggs

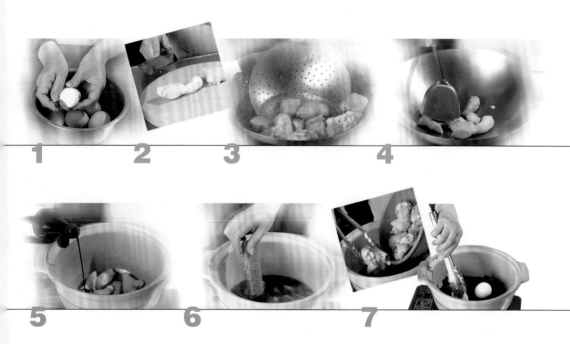

製法

1. 雞蛋焓熟浸凍水，去殼。
2. 薑洗淨刮去皮吹乾略拍。
3. 豬腳飛水過冷後，再以過面清水煮約1小時取出，過冷瀝乾備用。
4. 白鑊烘薑片刻。
5. 將薑放好於瓦鍋內，注入酸醋及清水，滾後以慢火煲1/2小時，熄火，原鍋浸過夜。
6. 次日加入片糖，煮至糖溶，再以慢火煲30分鐘。
7. 加入豬腳，再煲30分鐘，熄火焗至明天，再煲滾醋，以慢火再煮30分鐘，加入雞蛋，滾10分鐘後，熄火浸至入味。

Method

1. Hard boil eggs and soak in cold water. Peel.
2. Wash ginger, scrap off skin and air dry. Pound slightly.
3. Blanch pork knuckles in boiling water and rinse. Cover with water and boil for about 1 hour. Remove and rinse with cold water. Set aside.
4. Fry ginger slices in dry pan briefly.
5. Place ginger in a clay pot. Pour in vinegar and water. Bring to the boil and lower heat to simmer for 1/2 an hour. Turn off heat and allow to rest in pot overnight.
6. Add sugar slabs the next day. Boil until sugar melts and simmer for 30 minutes.
7. Add pork knuckles and simmer for a further 30 minutes. Turn off heat and rest until the following day. Bring vinegar back to the boil and simmer for another 30 minutes. Place in eggs and boil for 10 minutes. Turn off heat and leave to soak until the flavour is absorbed.

添丁雞酒

Chicken Wine for New Mothers

貼士 TIPS

傳統上，嬰兒出生後，十二朝會派雞酒。

以糯米酒做的味道較香甜，色澤也較深。

物料無須先醃，煮時亦不可加味，因為會使物料食味變酸。此食譜中介紹的酒與水之比例所製成的湯汁，清甜可口，吃時蘸醬油同吃十分美味。

Traditionally, chicken wine is given out 12 days after a baby is born.

Double distilled, triple distilled or glutinous rice wine may be used for chicken wine. Using glutinous rice wine gives a darker colour and better flavour.

There is no need to marinate ingredients or add seasonings, as this will make them go sour. The proportion of wine and water listed in the recipe gives a clear delicious soup. Dip in soy sauce when eating for added flavour.

材料	Ingredients
光雞（斬細件）1/2隻	1/2 dressed chicken (cut into small pieces)
豬肝（切片）2兩（80克）	2 taels (80g) pig's liver (sliced)
瘦肉（切片）2兩（80克）	2 taels (80g) lean pork (sliced)
木耳（浸透切絲）1/2兩（20克）	1/2 tael (20g) dried wood ears (soaked and shredded)
雙蒸酒或糯米酒1杯	1 cup double distilled white wine or glutinous rice wine
清水3杯（750毫升）	3 cups (750ml) water
雞雜（洗淨切塊）2副	2 sets chicken intestines (washed and cut)
雞蛋（打散）2隻	2 eggs (beaten)
花生仁（浸透）2兩（80克）	2 taels (80g) shelled peanuts (soaked)
薑絲1兩（40克）	1 tael (40g) shredded ginger
薑汁1/4杯	1/4 cup ginger juice

製法

1 光雞、雞雜、豬肝、瘦肉等，分別飛水，過冷瀝乾。
2 用油在鑊中炒雞蛋，盛起備用。
3 燒油爆薑絲，先將雞件、花生、木耳等加入炒透。
4 瓚酒，下薑汁及清水，待滾後改慢火煮20分鐘。
5 加入豬肝、瘦肉、雞雜及雞蛋，再煮10分鐘即成。

Method

1 Blanch dressed chicken, chicken intestines, pork liver and lean pork in boiling water separately. Rinse in cold water and drain.
2 Fry eggs in pan with oil. Dish and set aside.
3 Heat oil and fry ginger. Add chicken pieces, peanuts and wood ear. Fry until mixed.
4 Splash in wine and add ginger juice and water. Bring to the boil and simmer for 20 minutes.
5 Add pork liver, lean pork, chicken intestines and egg. Cook for a further 10 minutes.

鹵水鵝
Spiced Goose

貼士 | TIPS

此為綜合鹵味的菜式。綜合鹵味的意思，在於其鹵汁除了鹵製鴨鵝之外，還可以鹵其他物料，如牛腱、鴨腎、豆腐、雞蛋、雞翼及豬肉等。

鹵水盆用得越久，鹵出來的食品便會更香更可口，但應該注意衛生，以免滋生細菌；家庭製作，由於並非常用，最好把餘下的棄去，比較安全。

餘下之鹵汁，於再用時必須重新煮滾，並添加水分、調味及香料，否則鹵汁的食味會一次比一次淡。添加的各項用料，增量的比例必須平均。只加水分，味淡而質稀；只加調味，則味濃且杰身，此點應多加留意。

一鍋香醇的鹵水汁，只要你用浸、鹵或焗的方法來處理食材，使其慢慢入味，即可做出多種不同的鹵水食物來。

This is an all round spice sauce recipe. All round means that besides making goose or duck, it is suitable for other ingredients e.g. shin of beef, duck gizzard, bean curd, eggs, chicken wings and pork etc.

The longer the sauce is used the food produced from using it tastes even better. But attention must be paid to health and avoid the growth of bacteria. For home cooking, as it is not used regularly, it is better to discard after using.

The left over spice sauce must be re-boil before using again. Add more water, seasonings and spices. Otherwise the spiced food will taste more bland each time. When adding ingredients, they must be according to proportion. If only water is added the taste will be thin. If only seasonings is added it will be thick and strong. Attention should be paid to this.

With a pan of full flavour spice sauce, all you need to do is to soak, simmer or slow cook the food ingredients. It will gradually absorb the flavour and lots of different types of food can be produced.

材料

光鵝或米鴨1隻
（重約4斤）

鹵水料

八角6粒
甘草2片
草果1個
丁香1茶匙
沙薑6粒
陳皮1塊
桂皮2片

調味料

生抽1杯（250毫升）
鹽3湯匙
片糖3/4杯（切碎）
老抽1/4杯
玫瑰露酒1/4杯
清水12杯
味粉1茶匙

汁用料

蒜茸2粒
紅辣椒1隻（切幼絲）
白醋3湯匙
砂糖1茶匙

Ingredients

1 dressed goose or duck
(about 4 catties)

Spice Sauce

6 star aniseed
2 pcs Liquorice
1 grass nut
1 tsp cloves
6 dried stem ginger
1 pc dried tangerines peel
2 pcs cinnamon

Seasonings

1 cup (250ml) light soy
sauce
3 tbsp salt
3/4cup slab sugar (chopped)
1/4 cup dark soy sauce
1/4 cup rose wine
12 cups water
1 tsp MSG

Sauce

2 cloves garlic
1 red chilli (shredded)
3 tbsp white vinegar
1 tsp white sugar

1

2

3

4

製法

1 將鵝洗淨，去肺，放大滾水中汆水片刻
   過冷。
2 用大煲一個，加入鹵水料及調味，滾後用
   中火煮30分鐘。
3 放下已汆水之鵝，用慢火煮50分鐘，收火
   後再浸1小時。
4 取出候冷，斬件上碟，碟邊伴芫荽。（鹵水
   汁可留作鹵其他食物。）

Method

1 Wash goose, remove lungs and blanch in boiling
   water and then rinse in cold water.
2 Using a big pan, add spice sauce and seasonings.
   After boiling, lower heat to medium and simmer
   for 30 minutes.
3 Place in blanched goose. Cook on low heat for
   50 minutes. Turn off heat and soak for another
   hour.
4 Remove and when cold, chop into pieces
   and place on plate. Surround with coriander
   to garnish . (Save the spice sauce for other
   ingredients.)

三色千層糕
Tri-colour Thousand Layered Cake

材料

粘米粉1杯
木薯粉2杯
椰汁4杯（約1公升）
鹽1/4茶匙
糖霜1.5杯
紅、綠色素各適量

Ingredients

1 cup rice flour
2 cups tapioca flour
4 cups coconut milk (1L)
1/4 salt
1.5 cups icing sugar
some red & green food colouring

製法

1 將糖霜加入椰漿內與鹽一同攪至完全溶解。
2 再加入粘米粉及木薯粉，攪勻成粉漿。
3 將粉漿攪勻後分成三等份，其中兩份分別加入不同顏色素拌勻。
4 蒸籠內放上已塗油蒸熱之焗盆，倒入一層薄薄的粉漿，上蓋蒸3分鐘。
5 再倒入一層白色的粉漿蒸3分鐘，如此輪流鋪上不同顏色的粉漿，每層蒸3分鐘，鋪至最後一層，再蒸15分鐘，待冷卻後切成菱形上碟。

Method

1 Mix icing sugar, coconut milk and salt in a bowl until sugar is dissolved.
2 Sift in both rice flour and tapioca flour and stir into a batter.
3 Divide batter into 3 portions. Leave one of the remains in white while adding red and green food colourings to the other two portions respectively.
4 Heat up a greased cake mould in a steamer. Pour in a layer of coloured batter and steam for 3 minutes until it sets.
5 Add a white layer of batter and steam again for 3 minutes. Repeat the process of steaming layers of batter in alternate colours . After steaming the final layer, steam for another 15minutes. Leave it to cool, cut into diamond shape with a knife and serve.

材料

大菜約10克

糖約450克

麥芽糖約20克

罐裝栗子茸350克

清水1 3/4杯（190毫升）

Ingredients

10g agar agar

450g sugar

20g maltose

350g chestnut puree

1 3/4 cups water (190ml)

製法

1　大菜略洗瀝乾，用清水1 3/4杯浸約1小時備用。

2　將大菜連水煮至溶解，加入糖和麥芽糖等，以慢火煮至黏稠狀為合。

3　加入栗子茸攪拌均勻，然後倒進已抹油之平底糕盤中，置於冰箱中待凝固後，切件即可進食。

Method

1　Wash and drain the agar agar. Then soak it in 1 3/4 cups of water for 1 hour and set aside.

2　Bring agar agar with water to boil until dissolved. Add sugar and maltose to it and simmer until thickened.

3　Stir in chestnut puree and mix well. Pour the mixture into a greased baking tray and chill in refrigerator until solidified. Slice and serve cold.

粟一燒
Grilled Sweet Corn

材料
甜粟米4條

調味料
醬油適量
辣椒醬適量
熟油1湯匙

醃料漿
蜜糖1湯匙
麵粉2湯匙
清水適量

Ingredients
4 sweet corn

Seasonings
some thick soy sauce
some chilli paste
1 tbsp cooked oil

Marinade
1 tbsp honey
2 tbsp flour
some water

製 法
1 甜粟米洗淨，用水煮熟約20分鐘。
2 以竹籤穿在粟米的末端位置備用。
3 醃料預先調勻成流質狀，放入粟米沾上醃料漿，然後風乾。
4 將粟米放在烤爐上，以高火烤焗，其間要時常塗上調味料和熟油，直至粟米呈金黃色即成。

Method
1 Wash and the poach the sweetcorn in hot water for 20 minutes.
2 Pierce the sweetcorn with a skewer and set aside.
3 Mix the marinade into sauce. Dip the corn in the sauce and set aside until dry.
4 Arrange the sweetcorn on a grilling rack in an oven and roast over high heat. While roasting, brush the stirred seasonings onto sweetcorn frequently until they turn golden brown and serve.

材料
板豆腐1/2塊(100克)
糯米粉150克
清水適量

調味A(全部調勻煮濃稠)
紅糖5湯匙
醬油4湯匙
生粉少許
清水2杯(500毫升)

調味B(拌勻)
花生粉1/2杯
黃砂糖1/2杯

調味C
芝麻糖粉適量

Ingredients
1/2 piece (100g) beancurd
150g glutinous flour
some water

Seasonings A (mixed and thickened)
5 tbsp brown sugar
4 tbsp soy sauce
dash of cornflour
2 cups water (500ml)

Seasonings B (tossed)
1/2 cup peanut powder
1/2 cup brown sugar

Seasonings C
Some sesame sugar

製法

1. 豆腐及糯米粉放碗中,用手搓勻,使其充分混合至軟硬適中(如太硬可加少許水),搓成每個約12克的湯圓備用。
2. 將湯丸放入沸滾水中煮熟至浮起。
3. 撈出,轉放冰水中片刻,盛起。
4. 用竹籤串成3個一串。
5. 把串好的湯圓部分塗上調味A,部分灑上調味B及C,即可奉客。

Method

1. Put beancurd and glutinous flour in a mixing bowl and knead them into a dough by hands. Add some water if you find it too hard. Then make some glutinous balls of about 12g each with it.
2. Boil the glutinous balls in boiling water until they float.
3. Take out the balls, place them in the icy water for a short while and drain.
4. Link every 3 balls together with a bamboo skewer.
5. On the 3 balls in a skewer, paint seasonings A, sprinkle seasonings B and C respectively and serve.

中難度
Intermediate Pratice

```
                        麵點
                      Dim Sum
                         │
                         ▼
                      麵粉製品
                   Flour products
          ┌──────────────┼──────────────┐
          ▼              ▼              ▼
    油酥麵糰製品      膨脹麵糰製品      水調麵糰製品
 Pastry dough products  Risen dough products  Water dough products
          │              │              │
          ▼              ▼              ▼
```

- 混酥
  mixing pastry
- 包酥
  wrapping pastry
- 擘酥
  puffy pastry

- 生物膨鬆法製品
  biologically risen products
- 化學膨鬆法製品
  chemically risen products
- 物理膨鬆法製品
  physically risen products

- 冷水麵糰製品
  cold water dough products
- 熱水麵糰製品
  hot water dough products
- 溫水麵糰製品
  warm water dough products

註： 油酥麵糰因摻入是油脂，它不如水般容易滲透，許多時還是固體油脂，故調製時就要用雙手搓擦法，不能使用揉的方法，否則油脂與麵糰就不能完美結合成糰，這種雙手搓擦，稱為「擦酥」，需要的力度比較陰而強。

Remarks: For pastry dough, fat is used, which is not as osmotic as water, and usually even solid fat is used. Hence, rubbing with both hands instead of kneading should be applied in making the dough. Otherwise, the fat cannot combine with the dough well to form a lump. This method of using both hands to rub is called 'pastry rubbing'. The strength used should be soft and strong.

成形技法
Shaping Techniques

揉

揉即搓基本又簡單的成形技法，可分單手和雙手揉。（基礎法）

Kneading that is rubbing, the basic simple technique. You can knead by one hand or both hands. (Basic technique)

擀

擀運用麵棍、通心槌等工具將生坯製成不同形狀的主要技法，具有生坯成形與品種成形雙重作用。（基礎法）

Flattening a main technique to make the raw base into different shapes with the use of tools such as rolling pin and hammer. It serves the functions of both shaping the raw base and shaping the products. (Basic technique)

捲

捲將排好的坯料，經加餡、抹油或直接按製品要求捲成不同形狀如圓柱狀，並形成間隔層次的技法。分有單捲和雙捲。（基礎法）

Rolling a technique of rolling products into different shapes such as cylinder according to the requirements or after the base material is arranged, stuffed and brushed with oil. Spacing and layering effect is achieved. There are single rolling and double rolling. (Basic technique)

疊

疊經過擀製的坯料按需要經摺疊形成一定形狀的半製品之技法。（基礎法）

Folding a technique of folding flattened base material into certain shapes of semi-products according to needs. (Basic technique)

攤

攤將稀軟麵糰或糊漿入鍋或鐵板上，製成餅或皮的方法。特點是熟成形，意即邊成形邊成熟；或是用於稀軟麵糰或糊漿。它包括了旋攤、刮攤和手攤。（基礎法）

Spreading Transfer fluid dough or paste onto pot or metal plate to make pastry or skin. Its characteristic is done and shaped, that is forming shape during cooking. It is used in fluid dough or paste. It includes spin spreading, scrape spreading and hand spreading.

包

包將製好的皮子上餡後成形的技法，常用於提摺、燒賣、餛飩、湯包、春卷、糉子等。（基礎法）

Wrapping a technique of shaping the stuffed skin. It is commonly used in dumplings, pork and shrimp dumplings (siu mai), wontons, soup dumplings, spring rolls, rice dumplings, etc. (Basic technique)

捏

捏把包餡或不包餡的坯料按成品形狀的要求，運用手指技巧造型。主要分為一般捏法、捏塑法（提摺捏、推捏、捻捏、折捏、疊捏、扭捏、綜合捏等）。（基礎法）

Pinching Model the stuffed base or base without filling with fingers according to the shape requirements of the products. There are general pinching and model pinching (dumpling pinching, push pinching, twist pinching, bend pinching, fold pinching, turn pinching, integrated pinching, etc.)

剪 夾 按 抻

剪用剪刀把成品或半成品進行加工，讓它成形或便於成形的技法。（基礎法）

Trimming a technique of processing the products or semi-products with scissors to form the shape or help form the shape. (Basic technique)

夾借助工具如筷子、花鉗或花夾等將坯料夾製出一定形狀的技法。（基礎法）

Clipping a technique of clipping the base into certain shapes with the use of tools such as chopsticks, pattern tongs or clamps. (Basic technique)

按又稱為壓、掀、印，即用手將坯料印壓成形的技法。它可分為手掌按和手指按。（基礎法）

Pressing also called pushing, lifting, stamping; that is a technique of pressing the base into shape with hands. There are palm pressing and finger pressing. (Basic technique)

抻（伸）即拉麵，乃獨特的北方麵條技法。把柔軟麵糰經雙手反覆抖動、伸拉、扣合和折合成形。（北方技法）

Extending that is pulling noodles. It is a unique noodle technique in northern China. Vibrate the soft dough repeatedly with hands, extend and pull, and lock and bend to form the shape. (Northern technique)

模印

模印將生熟坯料注入、篩入或按入模具裏，利用模具成形的技法。它分有印模、套模、盒模和內模等。（模具輔助法）

Moulding Put raw or done base into a mould by injection or sieving in or pressing. It is a technique using moulds to form shapes. There are press mould, set mould, box mould and internal mould. (Mould-assisted method)

拔

拔用筷子將稀糊麵糰撥出兩頭尖而中間粗的條，立即把被拔出的成品直接投入鍋裏煮熟，這是借助加熱成熟後成形的獨特技法。（北方技法）

Pulling up Use chopsticks to make strips thick in the middle and sharp at the two ends from the fluid dough. Immediately put the strips into a pot directly to cook until done. It is a unique technique making use of heating and cooking to form the shape. (Northern technique)

削

削用刀直接一刀接一刀地削麵糰而成的長方形麵條的北方技法。（北方技法）

Peeling a northern technique of peeling rectangular noodles from the dough one by one directly with a knife. (Northern technique)

切

切以刀將加工後定型的坯料分割成形的技法。（北方技法）

Cutting a technique of dividing the processed and fixed base into shapes with a knife. (Northern technique)

鉗花 滾沾 鑲嵌 裱花

鉗花運用小工具整塑成品或半成品的技法。它依靠鉗花工具形狀的變化，使麵點形成多種形態或花紋，增強美觀外貌。（模具輔助法）

Patterning with tongs a technique of shaping products or semi-products with the use of small tools. It makes use of the different shapes of pattern tongs to make the dim sum into different forms or patterns to enhance the appearance. (Mould-assisted method)

滾沾將餡料加工成球形或小方塊後通過着水增加黏性，在粉料中滾動，使表面沾上多層粉料而成形的技法。（模具輔助法）

Rolling and sticking Process the filling into balls or squares, increase the stickiness by adding water, and roll the products in powder. It is a technique of forming the shape by sticking layers of powder onto the surface. (Mould-assisted method)

鑲嵌通過在坯料表面釀裝或內部填夾其他原料而達到美化成品，增調口味的技法。（裝飾法）

Inlaying a technique of putting other ingredients on the surface or in between of the base for decorating and flavouring. (Decorating method)

裱花（唧花）將裝飾有油膏或糖膏原料的布或紙袋，通過手指擠壓，使裝飾料均勻地從袋嘴流出，唧製出各式圖案或文字的技法。（裝飾法）

Piping a technique of applying pressure to a pastry bag, a cloth or paper bag containing decorating fat or sugar, with fingers to squeeze the decorating material through the tip of the pastry bag evenly to make different kinds of patterns or words. (Decorating method)

馬蹄糕
Water Chestnut Cake

材料
馬蹄 1/2 斤（320克）
馬蹄粉 1/2 斤（320克）
蔗糖 12 兩（480克）
豬油 1 湯匙
清水 5.5 杯

Ingredients
1/2 catty (320g) water chestnuts

1/2 catty (320g) water chestnut flour

12 taels (480g) cane sugar

1 tbsp lard

5.5 cups water

製法
1. 馬蹄去皮洗淨切小粒。
2. 馬蹄粉先用1.5杯清水開勻。
3. 再將餘下之4杯清水加入蔗糖同煮至糖溶。
4. 然後加入豬油及馬蹄粒，待滾片刻，立即熄火。
5. 待3分鐘後將馬蹄粉水攪勻，撞落煮滾之糖水中，快手攪勻成杰糊狀。
6. 倒入抹油之糕盆中，蒸約1/2小時便熟。
7. 凍後便可切塊，放熱油中煎至兩面金黃進食。

Method
1. Peel water chestnuts, wash and cut into small cubes.
2. Mix water chestnut flour with 1.5 cups of water.
3. Boil the remaining 4 cups of water with the cane sugar until melted.
4. Add lard and water chestnuts. Boil briefly and turn off heat.
5. Wait for 3 minutes and then quickly pour in mixed water chestnut solution into the boiled sugar solution. Mix quickly to form paste.
6. Pour into a greased cake tin and steam for 1/2 hour until cooked.
7. When cold, cut into slices and pan-fried in heat oil until golden brown on both sides before serving.

材料

片糖8兩（320克）
清水400毫升
紅豆4兩（160克）
糯米粉4兩（160克）
澄麵2兩（80克）
生油30毫升

Ingredients

320g cane sugar
400ml water
160g red bean
160g glutinous rice flour
80g wheat starch
30ml cooking oil

製法

1　紅豆預先焓脸至起沙。
2　將200毫升清水與糯米粉、澄麵拌勻成糯漿。
3　餘下之200毫升清水與糖同煮溶後，拌入煮脸之豆沙及生油。
4　將拌勻之糯漿撞入煮熱之豆沙糖水中，邊撞邊攪勻成糊狀。
5　糕盤或模型塗油，倒入糊漿，以高火蒸約1小時，即成。

Method

1　Boil red bean until it turns tender.
2　Mix the glutinous flour and wheat starch with 200ml water into a batter.
3　Melt the sugar with the remaining 200ml of water until dissolved. Toss with the cooked red bean and cooking oil.
4　Keep stirring the hot red bean sweet soup while pouring the glutinous batter to it.
5　Brush the cooking oil on the cake mould or mould. Pour in the red bean mixture and steam it over high heat for about an hour and serve.

79

材料

糯米粉250克
粘米粉250克
砂糖150克
清水150毫升
焓脸紅豆100克

餡料

豆沙100克

裝飾

紅棗適量
糖蓮子適量
甜核桃適量
車厘子適量

Ingredients

250g glutinous rice flour
250g rice flour
150g caster sugar
150ml water
100g boiled red bean

Filling

100g red bean paste

Garnishes

Some red dates
Some sweet lotus seeds
Some sweet walnuts
Some cherries

製法

1 將糯米粉、粘米粉及糖置於盤中,加入清水,用筷子拌勻成乾濕狀。
2 用手搓散,再用篩篩出細粉粒,然後拌入紅豆。
3 將糕盤放在蒸籠內,倒入一半有紅豆之粉粒,抹平,以大火蒸8分鐘。
4 將糕取出,平均鋪上豆沙,再把餘下之粉粒蓋在豆沙上,抹平。
5 將裝飾用之材料,適量排放在表面,再以大火蒸30分鐘即成。

Method

1 Put the glutinous rice flour, rice flour and sugar in a mixing bowl. Add in water and mix them up with a pair of chopsticks.
2 Knead it by hands, sift out the tiny flour particles and then mix with the red beans.
3 Put the cake mould in a steamer, pour in half of the red bean mixture, level its top and then steam it over high heat for 8 minutes.
4 Take out the cake, spread the red bean paste on it, cover it with the remaining red bean mixture and then level its top.
5 Place some garnishes on top and steam it over high heat for 30 minutes.

材料

綠豆粉 640 克
糕粉 80 克
白芝麻 60 克
北杏 5 克
南杏 80 克
清水 120 毫升
白糖 480 克
豬油 200 克
冰肉 120 克

Ingredients

640g green bean powder

80g cooked glutinous rice flour

60g sesame seeds

5g bitter almond

80g almond

120ml water

480g sugar

200g lard

120g fat pork

貼士 | TIPS

冰肉製法

肥肉約 160 克飛水過冷，切成薄片，再以少量砂糖拌勻，置冰箱，2 星期後備用。

Fat Pork Method

Blance 160g pork fat, rainse and slice. Add little sugar to stir well. Place in a refrigerator to chill for 2 weeks.

製法

1 先將芝麻、南北杏同炒香，放攪拌機中打碎。
2 綠豆粉和糕粉傾在檯上開穴，加入南北杏粉拌勻，倒入白糖和豬油，再注入清水拌至均勻，以乾散為好，切忌撥成一堆。
3 先將一半粉料放入餅印內，中心加上一片薄冰肉，然後再加上粉料，壓實印成餅型敲出。
4 放在焗盤上，以水壺噴上少許水。
5 入焗爐以 100℃ 爐火焗約 1 小時，而面火不可過高，焗至淺黃色為合。

Method

1 Parch sesame seeds and almonds until fragrant then grind well.
2 Sift green bean powder and cooked glutinous rice flour on a table, add sesame seeds and almonds to mix well. Make a hole in centre, put sugar and lard to mix together evenly. Add water to into the flour until the pastry is dry and separated.
3 Put half of the pastry in a cake mould and put in a piece of pork. Cover the pastry again. Press, then take the cakes out by knocking at the mould.
4 Arrange the cakes on the baking tray and sprinkle a little water on the cakes.
5 Bake the cakes in a preheated oven at 100°C for 1 hour until light brown.

狀元紅豆糕
Superior Red Bean Cake

貼士 TIPS

製作紅豆糕有兩種方法，其中一種是將紅豆煮腍，但保持原粒狀，加入粘米粉同蒸。用粘米粉製作的紅豆糕，蒸的時間較長，必須用上1小時15分以上，糕才能全熟。進食時，需要煎香，才會好吃。

今次介紹的狀元紅豆糕，蒸製時間，只須30分鐘，是以全豆沙製作，並改用馬蹄粉。馬蹄粉的好處，在於口感清爽彈牙，柔中帶韌，食時無須煎製，仍然滋味無窮。且甜味清淡，適合減肥人士。

There are two ways to make red bean cake. The one is used here to boil red beans until soft but retaining its original shape. Add to rice flour mixture as is and steam until cooked. By using rice flour to make red bean cake, the steaming time has to be extended to 1 hour 15 minutes or more. Pan fry the cake before serving to obtain its flavour.

The above recipe for red bean cake requires only 30 minutes of steaming time as the cake is made with pure red beans and water chestnut flour. The advantage of using water chestnut flour is that the texture of the cake is soft but chewy and non-sticky at the same time. It can be served without frying and still tasty. It is not too sweet and is ideal for people who are watching their weight.

材料

紅豆1/2斤(320克)	
馬蹄粉1/2斤(320克)	
黃砂糖2.5杯	
豬油1湯匙	
清水6杯(1500毫升)	

Ingredients

1/2 catty (320g) red beans
1/2 catty (320g) water chestnut flour
2.5 cups brown sugar
1 tbsp lard
6 cups (1500ml) water

1

2

3

4

5

6

製法

1 紅豆洗淨，加入清水8杯煮滾後，用慢火煮約1.5小時至開花鬆軟，水分收乾。
2 盛起待凍，壓成豆沙備用。
3 馬蹄粉先用1.5杯清水開勻。
4 將餘下之4.5杯水加入黃砂糖，煮至糖溶，然後加入豬油及紅豆沙，待滾後熄火，待涼3分鐘。
5 將馬蹄粉漿攪勻，撞落煮滾之糖水中，快手攪勻成杰糊，倒在抹油盆中，蒸1/2小時即成。凍後切件，放熱油中煎至金黃進食。

Method

1 Wash red beans and place in pan with 8 cups of water. When water comes to the boil, lower heat and simmer for about 1.5 hours until soft and tender and all the water has been absorbed.
2 Dish and when cold mash into puree and set aside.
3 Mix water chestnut flour with 1.5 cups of water.
4 Add brown sugar to the remaining 4.5 cups of water. Boil until sugar melts. Place in lard and red bean puree. When mixture comes back to the boil, take off heat and cool for 3 minutes.
5 Stir water chestnut flour solution and add quickly to sugar mixture. Stir until a thick paste is formed. Pour into greased cake tin and steam for 1/2 hour. When cold, cut into slices and fry in hot oil until golden brown on both sides and serve.

黃金南瓜糕
Golden Pumpkin Cake

貼士 | TIPS

因南瓜的甜味不一，壓茸後應先試味，如南瓜甜味足夠，可減去半杯糖。若不夠便要依足2杯半的份量，效果才會好。

蒸熟後，最好冷凍一天使其稍硬身後切件煎香進食，甘香軟糯，非常可口。如即日食用，略嫌軟身，口感欠佳。

As the sweetness of pumpkins varies, taste after the pumpkin has been mashed. If it is sweet enough, reduce the sugar by 1/2 a cup. If it is not too sweet, use the whole 2.5 cups sugar as listed in the ingredients for a fuller flavour.

After steaming, chill for at least a day to harden slightly. Before pan-frying, the cake will be soft and chewy. If it is served at the same day, the cake will be too soft and lack of in texture.

材料

南瓜1/2斤(320克)

粘米粉4兩(160克)

黃砂糖2.5杯

油1湯匙

清水2杯(500毫升)

Ingredients

1/2 catty (320g) pumpkin

4 taels (160g) rice flour

2.5 cups brown sugar

1 tbsp oil

2 cups (500ml) water

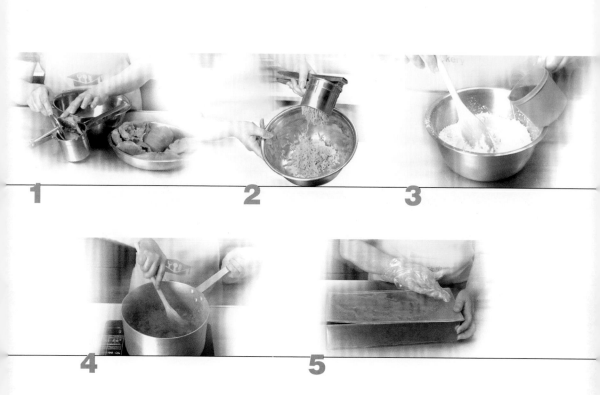

製法

1 南瓜去皮及籽，蒸熟壓成茸。

2 將粘米粉與清水1杯調勻。

3 餘下之清水加糖煮沸，將粉漿撞入，加油及南瓜茸，快手攪透。

4 倒入已抹油之盤(或模具)中，蒸1小時即成。

Method

1 Peel and remove seeds from pumpkin. Steam until cooked and mash.

2 Mix rice flour with 1 cup of water.

3 Place brown sugar with remaining water and bring to the boil. Add flour mixture, oil and mashed pumpkin. Stir quickly to mix.

4 Place in greased tin (or mould) and steam for 1 hour until cooked.

清香蜂蜜紫米糕

Honey Flavoured Black Glutinous Rice Cake

點士 TIPS

此為創新糕點，特點是可煎香熱食或雪凍冷食，兩者皆宜。

黑糯米亦稱赤糯或珍珠紅糯因其色較深，故一般多稱黑糯。黑糯含維他命B亦多，因其膠性重，不合用作主要食糧。

This is a new recipe and it is ideal to enjoy it hot after frying in oil or cold after chilling.

Black glutinous rice, is also called red glutinous rice or peal glutinous rice. The color is very dark so people usually name them black glutinous rice. The black glutinous rice consists of rich vitamin B but with strong glutin, whichmake it not suitable for our daily diets.

材料	調味料	Ingredients	Seasonings
黑糯米(紫米)4兩(160克)	高粱酒 1/2 杯	4 taels (160g) black glutinous rice	1/2 cup Lein wine
粘米粉 4 兩(160 克)	蜂蜜 1/2 杯	4 taels (160g) rice flour	1/2 cup honey
粟粉 4 兩(160 克)	椰糖 3 湯匙	4 taels (160g) corn flour	3 tbsp palm sugar
桂圓 2 兩(80 克)	油 1/2 湯匙	2 taels (80g) dried logan fruit	1/2 tbsp oil
花生粉 2 兩(80 克)		2 taels (80g) ground peanuts	
清水 12 兩(480 毫升)		12 taels (480ml) water	
熟油適量(塗糕盤用)		Oil for greasing tin	

1

2

3

4

5

製 法

1. 黑糯米洗淨用清水浸過夜，磨成粉漿160克。
2. 桂圓用清水浸片刻至軟身切碎。
3. 黑糯米漿、粘米粉及粟粉置盤中。
4. 加入調味、桂圓和清水攪勻。
5. 以旺火大滾水蒸1小時。待凍，撒上花生粉。

Method

1. Wash and soak black glutinous rice overnight. Ground into 160g powder.
2. Soak dried logan fruit until soft and chop finely.
3. Place ground black glutinous rice, rice flour and corn flour in bowl.
4. Add seasonings, dried logan fruit and water. Mix well.
5. Steam on high heat over boiling water for 1 hour. Allow to cool and sprinkle on ground peanuts.

貼士 TIPS

此點亦叫「蓮蓉糕」，是以蓮子焓腍壓至極爛成蓮蓉做成。但我卻喜愛入口有質感，所以壓蓮蓉時，略壓即可，吃時便會吃到碎粒的蓮子及香滑的蓮蓉。

可用份量一樣的粘米粉代替馬蹄粉，但馬蹄粉比較爽口清甜，且蒸的時間比粘米粉短，半小時便能完成，用粘米粉做便要蒸1小時以上。鏟蓮蓉時，時間控制十分重要，若時間過久，會使蓮蓉乾固而變成翻砂，製作便告失敗。

蓮蓉可預先做好，儲存冰箱中備用。

It is also called Mashed Lotus Seed Pudding which is made of finely mashed lotus seed paste. Since I love a bit chewy texture of the lotus seed, I just mashed it slightly to make a delicate paste with some chunks in it.

You may substitute water chestnut powder with rice flour. However, water chestnut powder is sweeter and takes shorter time for steaming, only about half an hour will do. On the other hand, it will take about more than an hour to steam the batter made of rice flour.

It is very important to control the timing for stir-frying the lotus seed paste. If it is too long, the paste will become so dry that it will tarnish the whole dish.

The paste can be prepared beforehand and stored in the refrigerator before use.

材料	蓮蓉材料	Ingredients	Ingredients for lotus seed paste
蓮子蓉 320克	白蓮子 320克	320g lotus seed paste	320g white lotus seed
馬蹄粉 320克	白糖 320克	320g water chestnut powder	320g sugar
黃砂糖 2.5杯	豬油 80克	2.5 cups brown sugar	80g lard
清水 6杯 (1500毫升)	滾水 2.5杯	6 cups (1500ml) water	2.5 cups boiled water
	鹼水 1湯匙		1 tbsp alkaline water

製法 Method

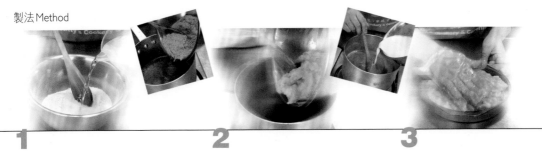

1　　**2**　　**3**

蓮蓉製法 Making aking of lotus seed paste

1　　**2**　　**3**　　**4**　　**5**

製法

1. 馬蹄粉先用 1.5杯清水開勻。
2. 將餘下之 4.5杯水加入黃砂糖，煮至糖溶，然後加入蓮子蓉，待滾後熄火，待涼 3分鐘。
3. 將馬蹄粉漿攪勻，撞入煮滾之糖水中，快手攪勻成稠糊狀，倒在已抹油之盆中，蒸 1/2小時即成。凍後切件，放熱油中煎至金黃進食。

蓮蓉製法

1. 蓮子用鹼水拌勻，以濕透為度，約 15分鐘沖水。
2. 倒入滾水內煮 10分鐘，使它滾至均勻。
3. 取出過冷挑去蓮芯，洗淨。
4. 將蓮子放煲中，加水至過面，煮至夠腍為準，取出略為壓爛。
5. 將蓮子蓉、白糖和豬油放下鑊中，不停鏟動至粘稠即成。

Method

1. Mix the water chestnut powder in 1.5 cups of water.
2. Melt the brown sugar with the remaining 4.5 cups of water. Add the lotus seed paste and bring to a boil. Turn off the heat and leave it for 3 minutes.
3. Stir the water chestnut powder mixture thoroughly. Pour it into the boiled sweet soup. Mix it quickly into a batter by hands. Pour it into an oil-brushed container. Steam it for 1/2 hour. Cut it into pieces when cool. Deep-fry it in hot oil until golden brown before serving.

Making aking of lotus seed paste

1. Mix the lotus seed with alkaline water and have it soaked for about 15 minutes before rinsing.
2. Pout it into hot water and cook for 10 minutes.
3. Refresh the seed, remove its core inside and wash.
4. Put the lotus seed in a pot. Add in enough water to cover the seed. Boil until it turned tender. Then take it out and mash.
5. Put the lotus seed paste, sugar and lard in a wok and saute continuously until it thickened and serve.

貼士｜TIPS

軟滑蘿蔔糕多在酒樓食肆才可吃到，與蘿蔔糕的做法差不多，分別在於前者用的蘿蔔及水分較多，但粉卻較少；質地比較腍，不宜煎食，吃時要用匙舀取，以熱食為佳。

若一天吃不完，可置於雪櫃中，吃時再蒸熱即可。

此軟滑蘿蔔糕，必需要熱食，不能煎。

Soft turnip cake is normally available in the restaurants. The cooking method is similar to the standard turnip cake. The difference is that there are more turnip and water occupied with less flour. The texture is softer and is not suitable for frying. It is better to serve hot with spoon.

If they cannot be finished in a day, place in refrigerator and steam to warm before serving.

This soft version of turnip cake has to be eaten hot and is not suitable for frying.

材料

白蘿蔔2.5斤（1600克）
臘腸1條
臘肉1/2條
蝦米1兩（40克）剁碎
冬菇1/2兩（20克）浸透切粒
清水2.5杯（625毫升）
芹菜1棵（切粒）
葱2條（切粒）
蒜茸1茶匙

粘米粉2兩（80克）
鷹粟粉2兩（80克）

調味料

糖1茶匙
味粉1茶匙
豬油2湯匙
鹽1茶匙
胡椒粉1/2茶匙

Ingredients

2.5catties (1600g) white turnip

1 pc Chinese sausage

1/2 strip preserved belly of pork

1 tael (40g) dried shrimps chopped

1/2 tael (20g) mushrooms, soaked and diced

2.5 cups (625ml) water

1 stick Chinese celery (diced)

2 pcs spring onions (diced)

1 tsp crushed garlic

2 taels (80g) rice flour

2 taels (80g) corn flour

Seasonings

1 tsp sugar

1 tsp MSG

2 tbsp lard

1 tsp salt

1/2 tsp pepper

製 法

1. 蘿蔔去皮切成絲。
2. 以1杯清水置煲中煮滾，再轉慢火煮至軟身，但尚未變色為合。
3. 臘味及各配料均切粒，以1粒蒜茸起鑊爆香盛起備用。
4. 粘米粉及粟粉用1.5杯清水開勻，即可加入蘿蔔中，趁熱攪成稀糊，然後再加入爆香之臘味、冬菇、蝦米及調味攪勻。
5. 將糕盆或小碗掃油，把粉漿分放其中，放在蒸籠上，蒸約30分鐘便熟，取出前撒下芹菜粒及葱粒。
6. 食前撒少許胡椒粉及熟油、醬油。

Method

1. Peel white turnip and shred.
2. Boil 1 cup of water in pan and simmer until soft but still retains the colour.
3. Dice preserved meat and other ingredients. Fry in wok with crushed garlic until cooked. Set aside.
4. Mix rice flour and corn flour with 1.5 cups of water. Add to turnip. Stir while hot to form thin paste. Add the fried preserved meat, mushrooms, dried shrimps and seasonings. Mix well.
5. Grease pan or little bowl with oil. Pour flour mixture in these containers and place in steamer. Steam for about 30 minutes until cooked. Sprinkle on chopped celery and spring onion before removing.
6. Before serving sprinkle on pepper, cooked oil and soy sauce.

臘味蘿蔔糕
Turnip Cake with Preserved Meat

貼士 | TIPS

水分和粉料的比例能否配合，是影響蘿蔔糕之口感的重要因素。這裏所用之份量，應以3杯水（包括煮蘿蔔水）為準，多餘的水分不要。倘若煮水不足3杯，必須加清水至3杯為合。

蘿蔔煮好後，應趁熱拌入粉漿，使成杰糊狀。如放置過久，熱力減退才製作，漿料便會不均勻，糕可能會出現半生熟狀態，應加注意。

用筷子試插，是為保險之做法，若仍有粉漿黏着，便應多蒸約15分鐘為宜。

一般家庭皆用刨刀把蘿蔔刨成絲。此處建議用刀切的方法，能有效地增加口感。蒸熟後煎至金黃的蘿蔔糕，入口軟硬適中，甘香而不乾涸，滲發着濃郁的蘿蔔汁液，食味誘人。

Portion of water and ingredient is an important factor, which affects the texture of the cake. The quantity used here i.e. 3 cups of water (including water in which the turnip is cooked) is a standard, discard excess water. If the cooking water is less than 3 cups, add more water to make up the difference.

After the turnip is cooked, add the remaining ingredients while still hot in order to form a paste. If it is left for too long and the mixture will get cold, and cannot be evenly spread through. Final product may be semi-cooked

Requiring 1 hour of steaming, according to the quantity mentioned above. Testing with a chopstick is a safety way. If the mixture is still sticky, let it steam for another 15 minutes.

It is a common use a peeler to shred turnips. Here the turnip is cut into shreds, which improves the texture. When the cake is cooked, pan fried until golden brown. Finishing is neither too soft nor too hard, tasty and non sticky. It has a strong turnip flavour and very mouth watering

材料	調味料	Ingredients	Seasonings
蝦米1兩(40克)	鹽1湯匙	1 tael (40g) dried shrimps	1 tbsp salt
粘米粉1斤(640克)	豬油2湯匙	1 catty (640g) rice flour	2 tbsp lard
蘿蔔4斤(2560克)	味粉2茶匙	4 catties (2560g) white turnip	2 tsp MSG
白芝麻1湯匙(炒香)	胡椒粉適量	1 tbsp white sesame seeds (fried)	pepper to taste
臘腸2條		2 Chinese sausages	
葱1條(切粒)		1 spring onion (diced)	
冬菇1兩(40克)		1 tael (40g) mushroom	
清水3杯(750毫升)		3 cups water (750ml)	

製法

1. 蝦米洗淨浸透、冬菇浸透切粒、臘腸切粒。
2. 蘿蔔去皮洗淨切幼條。
3. 燒鑊加油爆香蝦米、冬菇、臘肉等材料，加入調味兜勻盛起。
4. 用2杯水同煮至蘿蔔軟身盛起。
5. 將煮蘿蔔之水，量出3杯份量。
6. 粘米粉倒入開勻，再把所有材料及調味加入，趁熱攪成杰糊。
7. 糕盆塗油。
8. 倒入混合料扒平，猛火蒸1小時，以筷子試插，若不黏米漿，即表示糕已蒸熟。
9. 取出撒上青葱粒及芝麻，待凍才可切件煎食。

Method

1. Wash and soak dried shrimps; soak mushrooms and dice; dice Chinese sausages.
2. Wash white turnip, peel and shred.
3. Heat wok and add oil. Fry dried shrimps, mushrooms and Chinese sausages. Add seasonings, mix and dish.
4. Boil turnip in 2 cups of water until soft. Dish.
5. Measure 3 cups of liquid from the turnip boiling mixture.
6. Add rice flour and mix well. Place in remaining ingredients and stir while hot to form paste.
7. Grease cake tin with oil.
8. Pour in mixture and smooth top. Steam on high heat for 1 hour. Test with chopstick. If it does not stick, the cake is cooked.
9. Remove and sprinkle on spring onion and sesame seeds. Allow to cool. Cut into slices and pan fry before serving.

貼士 TIPS

由於眉豆糕較挺身，蒸好後趁熱切件供食，不會鬆散。

此糕為豆製品，故與五香粉之食味十分匹配。喜愛五香味道者，可於炒製配料時加入，或直接拌入粉漿之中。不愛者則免。

中醫稱眉豆性溫，功效理中益氣，補腎健脾，如加入蓮藕煲湯佐膳，更有補血潤腸通便之功效。

As the black-eyed bean cake is quite firm, it can be sliced and serve while warm. The cake will not fall apart.

As this is a bean cake, it goes very well with five-spice powder. For those who like the taste of five-spice powder, add some while frying the ingredients or straightly stir into flour mixture. This can be omitted for those who prefer not to use it.

According to the advice from traditional Chinese medicine doctors, the black-eyed bean is good for throat and kidney. Adding some lotus root into the soup would help to expedite defecation and improve blood circulation.

材料	調味料	Ingredients	Seasonings
眉豆 1/2 斤（320克）	鹽 1/2 茶匙	1/2 catty (320g) black-eyed beans	1/2 tsp salt
粘米粉 5 兩（200克）	糖 2 茶匙	5 taels (200g) rice flour	2 tsp sugar
澄麵 1 兩（40克）	雞粉 1 湯匙	1 tael (40g) tang flour	1 tbsp chicken powder
清水 3.5 杯（875毫升）	胡椒粉 1/2 茶匙	3.5 cups (875ml) water	1/2 tsp pepper
臘腸 2 條（切粒）	豬油 2 湯匙	2 pcs Chinese sausages (diced)	2 tbsp lard
葱 2 條（切粒）		2 pcs spring onion (diced)	
蝦米 1 兩（40克）		1 tael (40g) dried shrimps	

製 法

1. 眉豆洗淨，以5杯清水煮腍（約1小時）。將豆盛起，煮豆之水倒去。
2. 用油爆香臘味、蝦米，加入調味兜勻盛起。
3. 用1杯清水開勻粘米粉及澄麵。
4. 將2.5杯清水置煲中，煮沸，撞入開勻之粉漿，邊撞邊攪勻至杰身。
5. 再把爆香之臘味及眉豆等加入攪透。
6. 糕盆塗油，將以上材料倒入盤中抹平。
7. 放蒸籠上，以大火大滾水蒸1小時，用筷子試插，若不黏米漿即成。

Method

1. Wash black-eyed beans and boil in 5 cups of water until soft (about 1 hour). Remove beans and discard water.
2. Fry preserved meat and dried shrimps in oil until fragrant. Mix in seasonings and dish.
3. Mix rice flour and tang flour with 1 cup of water.
4. Place 2.5 cups of water in pan and bring to the boil. Add flour solution, pouring and mixing all the time, until thickens.
5. Add fried preserved meat and black-eyed beans etc. and mix well.
6. Pour above mixture in greased cake tin. Smooth the top.
7. Place in steamer and steam on high heat over boiling water for 1 hour. Test with chopstick. If it does not stick, the cake is cooked.

芋頭糕
Taro Cake

必須留意：五香粉可説是芋頭糕之靈魂，倘若省去，注定不會好吃。

無論如何愛吃，芋頭的用量都不能太多，1斤足夠，過多會影響糕身硬度，不夠柔軟。

以往，長輩們製作芋頭糕時，愛加進臘鴨及蘇薑（即紅薑），除能突顯芋頭之食味之外，臘鴨尚帶增香之功；蘇薑則具消滯之效，是極佳的配搭。有興趣者不妨一試。

Special attention must be paid to the five spice powder as this is the soul of the taro cake. If it is omitted, the taro cake definitely will not taste good.

Even if a stronger taro flavour is preferred, the amount of taro used must not exceed 1 catty. Furthermore more will affect the cake too harden.

In the old days, people like adding preserved duck and red pickled ginger in the ingredients. To enhance the taste of taro and the aroma. And the red pickle ginger helps digestion. This is a good combination for those who interested should give it a try.

材料	調味料	Ingredients	Seasonings
荔甫芋1斤（640克）	五香粉1茶匙	1 catty (640g) laipo taro	1 tsp five spice powder
蝦米1兩（40克）	胡椒粉1茶匙	1 tael (40g) dried shrimps	1 tsp pepper
臘腸2條	鹽1湯匙	2 pcs Chinese sausage	1 tbsp salt
冬菇4隻	糖1.5茶匙	4 mushrooms	1.5 tsp sugar
粘米粉10兩（400克）	味粉1茶匙	10 taels (400g) rice flour	1 tsp MSG
澄麵2兩（80克）	麻油1茶匙	2 taels (80g) tang flour	1 tsp sesame oil
清水6杯（1500毫升）		6 cups (1500ml) water	
蒜頭1粒（略拍）		1 clove garlic (slightly crushed)	

1

2

3

4   **5**   **6**

製 法

1 芋頭去皮切粒。
2 將各材料切粒，用蒜頭起鑊爆香棄去，倒下各材料爆香，加入調味及芋頭兜勻盛起。
3 將4杯清水煮沸熄火。
4 另2杯凍水開粉漿，然後撞落熱水中，邊撞邊攪透成杰糊，加入已爆好之材料攪透。
5 糕盤塗油，將糕料倒落盤中，用手抹油拍打至平滑。
6 放入蒸籠以大火蒸1小時，可用竹筷試插，不黏米漿即表示芋頭糕已蒸熟。

Method

1 Peel taro and dice.
2 Dice remaining ingredients. Fry garlic in oil and discard. Add all ingredients to the oil and fry until fragrant. Add seasonings and diced taro. Mix and dish.
3 Bring 4 cups of water to the boil and turn off heat.
4 Using the remaining 2 cups of water to make a flour paste. Pour into boiled water, stirring and pouring at the same time, until mixture becomes thick. Add fried ingredients and mix well.
5 Grease cake tin and place in mixture. Grease hand and wipe top of cake until smooth.
6 Place in steamer and steam on high heat over boiling water for 1 hour. Test with Chopstice. If it does not stick, the cake is done.

貼士 TIPS

這是簡易年糕的做法，粉漿稀釋易熟。懷舊式的年糕，製作繁複，需蒸4小時以上，並要分多次攪動米漿。

蒸的時間長短，因模具之大小各異。小模蒸30分鐘即可，大模則要蒸上1小時 或1小時15分鐘。

以竹籤插入粉漿中，不黏者即熟。略凍後，於四邊稍撥，即可起出。

變化之一：「鴻運當頭小白龍」先注入紅色於魚頭部分，待2分鐘快將凝固時，即加入白色粉漿。

變化之二：「黃金錦鯉賀新禧」先將紅色粉漿注入魚模至近尾部，合蓋蒸約10分鐘，再注入白色粉漿。

如希望錦鯉顏色鮮明而帶半透明感，可於落粉漿後，約10分鐘，當粉漿尚未完全凝固時，滴入不同色漿即可。

This is a simple way to make New Year cake as the thin mixture is easy to cook. The old method of making cake is complicated and requires over 4 hours of steaming and constant stirring.

The steaming time required depends on the size of the fish mould. Small moulds take about 30 minutes and large mould may take about 1 hour to 1 hour 15 minutes.

Insert bamboo skewer into mixture and if it does not stick, the cake is cooked. Allow to cool slightly and gently loosen sides for the cake to slide out.

Variation 1: good Luck White Dragon?Place red mixture into head part and steam for 2 minutes until just set before adding white mixture.

Variation 2: golden Carp Celebrating New Year?Place red mixture into fish mould at the tail end and steam covered for 10 minutes before adding white mixture.

If a more transparent and colourful look is required, place mixture and steam for 10 minutes. Add various coloured mixture before the original mixture is completely set.

材料	Ingredients
白糖10兩（400克）	10 taels (400g) white sugar
清水13兩（520毫升）	13 taels (520ml) water
糯米粉5兩（200克）	5 taels (200g) glutinous rice flour
澄麵4兩（160克）	4 taels (160g) tang flour
椰汁2兩（80克）	2 taels (80g) coconut milk
生油1.5兩（60克）	60g oil
色素適量	dash of food colouring

1

2

3

4

製法

1. 清水與糖同煮溶，待凍。
2. 除色素外，將所有材料加入凍糖水中攪勻，然後取出1杯（約250克）加入色素攪勻。
3. 魚模塗油，放於蒸鍋內，先入適量色漿於頭、翅及尾部分，合蓋蒸3分鐘。
4. 即可加入白色粉漿，以高火蒸30分鐘即成。

Method

1. Boil sugar and water until melted, allow to cool.
2. With the exception of food colouring, place all ingredients in cold sugar solution and mix well. Pour out 1 cup (about 250g) of mixture and stir in suitable amount of food colouring.
3. Brush fish mould with oil and place in steamer. Place suitable amount of coloured mixture into the head, fins and tail of mould. Cover and steam for 3 minutes.
4. Add white mixture and steam on high heat for 30 minutes.

蝦米腸粉
Dried Shrimp Rice Sheet Rolls

材料
粘米粉1/2斤(320克)
清水1斤(640毫升)
熟油1.5兩(60克)

餡料
蝦米1兩(40克)，切粒
葱3條，切粒

調味料
鹽1茶匙
糖1/2茶匙

甜豉油
醬油3湯匙
味粉1/4茶匙
糖1/2茶匙
清水3湯匙

Ingredients

1/2 catty (320g) rice flour

I catty (640ml) water

1.5 taels (60g) cooked oil

Filling

I tael (40g) dried shrimps

3 pcs Spring onion, both diced

Seasonings

I tsp salt

1/2 tsp sugar

Sweet Soy Sauce

3 tbsp say sauce

1/4 tsp MSG

1/2 tsp sugar

3 tbsp water

製 法

1　粘米粉置盤中，加入熟油拌勻，使粉末充分吸入熟油約15分鐘。
2　加入清水1斤攪勻(邊加邊攪透)至粉末與水完全溶解。
3　再以篩隔淨，加入調味拌勻備用。
4　將蒸粉器放在已注水之鑊上燒滾，鋪上一幅濕布。
5　倒入米漿，撒下適量餡料，蓋密蒸3-4分鐘。
6　取出把粉及布反轉覆蓋，把布拉出，捲成腸粉。

Method

1　Place rice flour in bowl and mix with cooked oil. Leave for about 15 minutes to allow the flour to absorb the oil completely.
2　Add I catty of water and mix (stirring while pouring). Stir until the flour is completely dissolved.
3　Pass through sieve to remove impurities. Mix with seasonings and set aside.
4　Heat rice roll steaming sheet on top of a wok with boiling water. Cover with a wet cloth.
5　Pour in rice solution and scatter on some filling. Cover and steam for 3-4 minutes.
6　Remove and place cloth and cooked rice sheet upside down. Pull away cloth and roll into a rice sheet roll.

貼 士 | TIPS

甜豉油製法
將材料拌勻，煮滾待冷備用。

Method for Sweet Soy Sauce
Mix all ingredients boil and allow to cool.

甜腸粉
Sweet Rice Rolls

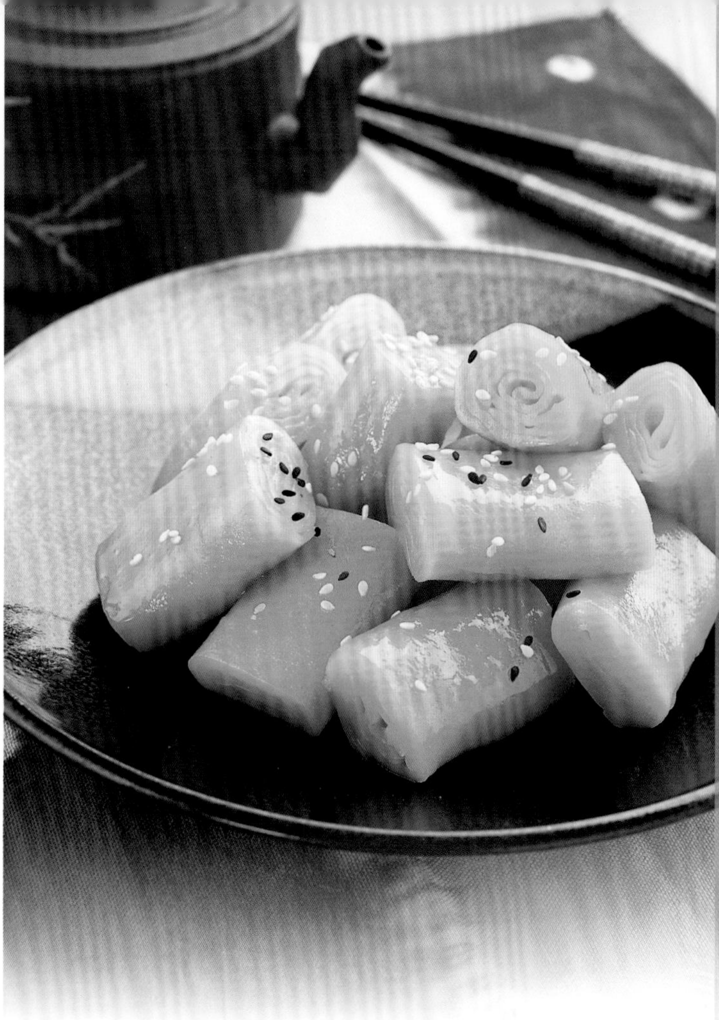

材料A

黃砂糖4兩(160克)
清水1/2杯(125毫升)

材料B

粘米粉4兩(160克)
粟粉2湯匙
清水2杯(500毫升)
油2湯匙
炒香芝麻少許

Ingredients A

160g brown sugar
1/2 cup (125ml) water

Ingredients B

160g rice flour
2 tbsp corn flour
2 cups (500ml) water
2 tbsp oil
Some toasted sesame

製法

1. 將材料A置煲中煮溶待凍。
2. 材料B開勻，與材料A混合拌透，再加入油拌勻。
3. 將蒸腸粉器放在已注水之鑊上燒滾，鋪上濕布或牛油紙。
4. 倒入米漿1湯勺，蓋密蒸3-4分鐘。
5. 把已蒸熟的粉漿連布一起取出並反轉覆蓋，然後把布拉出，捲成腸粉，灑上芝麻即成。

Method

1. Melt ingredients A in a pot. Then let it cool down.
2. Mix ingredients B well, blend it with the syrup from ingredients A. Add in the oil and stir.
3. Put the steamer in a wok of boiling water and place a wet cloth or baking paper on it.
4. Pour a tbsp of rice batter on the steamer, cover the lid tightly and steam it for 3-4 minutes.
5. Take out the cooked rice sheet together with the wet cloth, flip over, pull out the cloth and roll up the rice sheet. Dust it with some sesame and serve.

雞仔餅
Chicken Shaped Sweet Cake

皮料
麵粉 3 兩（120 克）
麥芽糖 2 兩（80 克）
蘇打食粉 1/2 茶匙
生油 1 湯匙
清水 2-3 湯匙

餡料
糖肥肉 4 兩（160 克）
砂糖 5 兩（200 克）
核桃肉 1 兩（40 克）
杏仁 1 兩（40 克）
欖仁 1/2 兩（20 克）
糕粉 4 兩（160 克）
芝麻 1 湯匙
豬油 1 兩（40 克）
清水 1/4 杯

調味料
鹽 1/2 茶匙
南乳 1 湯匙
紹酒 1 茶匙

Wrapping Ingredients
3 taels (120g) flour
2 taels (80g) malt
1/2 tsp bicarbonate of soda
1 tbsp oil
2-3 tbsp water

Filling
4 taels (160g) sugared pork fat
5 taels (200g) sugar
1 tael (40g) shelled walnuts
1 tael (40g) almonds
1/2 tael (20g) olive kernels
4 taels (160g) cooked glutinous flour
1 tbsp sesame
1 tael (40g) lard
1/4 cup water

Seasonings
1/2 tsp salt
1 tbsp preserved taro curd
1 tsp Shao Hsing wine

製法
1. 將餡料所有材料切成細粒，如白豆般大小粒狀。
2. 再把糕粉及調味等加入混和搓透，然後分成30份，每份搓成圓球形。
3. 麵粉與麥芽糖、蘇打粉等混和搓透成粉糰，再分成30份，每份開薄包入餡料，用手壓實按扁，排放烤盤中，表面塗上蛋液。
4. 將做好的雞仔餅放進焗爐中，以180℃焗15分鐘即成。

Method
1. Dice all filling ingredients, size of a white bean.
2. Add cake flour and seasonings. Mix and knead well. Divide into 30 parts and roll each part into a ball.
3. Mix flour with malt and bicarbonate of soda. Knead to form dough. Divide into 30 parts. Wrap each with filling and press to flatten. Place on baking tray and brush tops with beaten egg.
4. Place made chicken shaped cakes into oven and bake on 180°C degrees for 15 minutes.

材料

麵粉8兩（320克）
糖漿5兩（200克）
生油 1/2杯
鹼水 1/2茶匙

Ingredients

320g plain flour
200g golden syrup
1/2 cup oil
1/2 tsp alkaline water

製法

1　麵粉篩在盤中。
2　加入糖漿、生油及鹼水拌勻，搓成一軟糰放置3小時後方可取用，否則餅不易敲出。
3　將粉糰搓成長條，分成所需等份，搓圓。
4　將粉糰放入已灑粉之模形中按實。
5　將模向左右一敲取出，排放焗盤中，以190℃先焗2分鐘，取出塗上蛋液，再放入焗爐中，焗20分鐘，至表面金黃，取出待凍即成。

Method

1　Sift the flour into a mixing bowl.
2　Add the golden syrup, oil and alkaline water to it and knead into a soft dough. Leave it for 3 hours before use.
3　Knead the dough into a long stick. Divide it into several portions and knead them into some balls.
4　Put the dough ball into a mould sprinkled with flour. Press it hard.
5　Hit the left and right hand sides of the mould to take out the moulded dough. Line them on a baking tray. Bake it in 190°C for 2 minutes. Take it out and brush some beaten egg on the dough. Return it to the oven and bake for another 20 minutes until the cakes turned golden brown. Remove and leave to cool and serve.

眉豆茶粿
Black-eyed Bean Glutinous Bun

餡料可做包子或糯米糍。

皮料與紅龜粿的搓法及製法相同。

若即日吃不完,可置放雪櫃中,於進食前蒸5-10分鐘便可。

Filling can be used for buns or dumplings.

The wrapping is made the same way as for red turtle shaped glutinous buns.

If they are not finished in a day, store in refrigerator and steam for 5-10 minutes before serving.

材料	餡料	Ingredients	Filling
糯米粉5兩（200克）	眉豆1/2斤（320克）	200g glutinous rice flour	1/2 catty (320g) black-
粘米粉2.5兩（100克）	蝦米2湯匙（剁碎）	100g rice flour	eyed beans
鹽1茶匙	五香粉1茶匙	1 tsp salt	2 tbsp dried shrimps (minced)
油2湯匙		2 tbsp oil	1 tsp five spice powder
滾水7安士（200毫升）	調味料	7 oz (200ml) boiling water	
糉葉1塊	鹽1/2茶匙	1 pc bamboo leaf	Seasonings
	糖2茶匙		1/2 tsp salt
	雞粉1茶匙		2 tsp sugar
	胡椒粉少許		1 tsp chicken powder
			Dash of pepper

餡料製法 Method for Filling

皮製法 Method for Wrapping

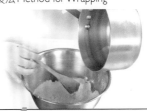

餡料製法

1 眉豆洗淨，以5杯清水煮腍（約1小時），隔去水分搓爛。
2 燒油爆香蝦米及五香粉，傾下搓爛之眉豆及調味，拌勻搓透成餡料。

皮製法

1 將兩種米粉與鹽同篩，置盤中加油及撞入滾水，以木匙攪拌。
2 稍涼即用手搓勻成軟滑之米糰。
3 將米糰分成12等份，搓圓按扁，包入適量餡料，收口處放上一小塊糉葉，置蒸籠以大火蒸10分鐘即成。

Method for Filling

1 Wash black-eyed beans and boil in 5 cups of water until soft (about 1 hour). Drain and mash.
2 Heat oil and fry dried shrimps and five spice powder. Add mashed black-eyed beans and seasonings. Mix and knead to form filling.

Method for Wrapping

1 Sieve together the two types of flour and salt. Place in bowl and add oil. Pour in boiling water and stir with wooden rod.
2 Cool slightly and knead with hands to form soft and smooth dough.
3 Divide rice dough into 12 equal portions. Roll into balls and press flat. Wrap in filling and place a piece of dumpling leaf at seam. Place in steamer and steam on high heat for 10 minutes.

1 皮料與小籠包相同，以生熟粉混合製作的好處，在於搓製時麵糰不會黏檯及麵棒，且口感清爽。

2 單煎一面的叫做「貼」。

3 鍋貼煎熟排放碟中，煎成金黃的底部向上放因這才是正面。

The wrapping is the same as for mini steamer buns. The advantage of using a raw and cooked dough combination is that it will not stick to the table or rolling pin. The texture when eaten is not sticky as well.

Only one side fried dumpling is called "tip".

When the dumplings have been cooked, please turn over them because the bottom of the dumplings becomes the top when putting on plate to serve.

A材料

麵粉 1兩(40克)	榨菜 1/2兩(20克)
清水 20毫升	薑米 1茶匙

B材料

	調味料
豬油 1/2兩(20克)	醬油 1湯匙
麵粉 4兩(160克)	蠔油 1湯匙
滾水 2兩(80克)	油 1茶匙
	麻油少許
餡料	胡椒粉少許
紹菜 4兩(160克)	鹽 1/2茶匙
豬肉(攪爛) 6兩	糖 1/2茶匙
(240克)	雞粉 1茶匙
馬蹄肉 1/2兩(20克)	生粉 1.5湯匙

Ingredients A

40g flour	20g Sichuan preserved vegetables
20ml water	1 tsp chopped ginger

Ingredients B

	Seasonings
20g lard	1 tbsp soy sauce
160g flour	1 tbsp oyster sauce
80g boiling water	1 tsp oil, dash of sesame oil
	Dash of pepper
Filling	1/2 tsp salt
160g Chinese cabbage	1/2 tsp sugar
240g minced pork	1 tsp chicken powder
20g water chestnuts	1.5 tbsp bean flour

餡料製法 Method for Filling

1 2 3 4 5

餡料製法

紹菜洗淨飛水，瀝乾水分後剁碎，與豬肉混合加入調味攪透，放雪櫃備用。

製法

1. 先將A料搓成麵糰，待發1小時後備用(此為生麵糰)。
2. 油與水煮滾，撞落粉中攪勻搓好。再與生麵糰混合搓成光滑軟糰。
3. 分成30粒粉子，以木棍開成圓薄形。
4. 包上餡料(約15克)成鍋貼形。
5. 煎鍋燒熱，塗油少許，排上鍋貼，加入清水4-5湯匙，蓋上鍋蓋以慢火煮10分鐘，待水乾加入少許油，以慢火煎至金黃即成。

Method for Filling

Wash vegetables and blanch in boiling water. Squeeze dry and mince. Mix with pork and seasonings and stir well. Pace in refrigerator.

Method

1. Mix ingredients A into a dough and set aside for 1 hour for it to rise (this is the raw dough).
2. Boil oil and water. Pour immediately into flour. Stir and mix well. Add to raw dough and knead into a smooth soft dough.
3. Divide into 30 pieces and with rolling pin, press into thin circles.
4. Wrap in ingredients (about 15g).
5. Heat frying pan and grease. Line potstickers in pan and add 4-5 tablespoons of water. Cover and cook slowly for 10 minutes. When all the water is gone, add a little oil. Fry on low heat until golden brown.

貼士 TIPS

包糉時若把糉子握得太緊，會把糯米滲進豆沙中，這樣豆沙周邊的米便蒸不熟。包上威化紙是很好保險的方法；也可以先把豆沙置冰格中雪硬才包。

包時要鬆手，能搖響為合，這樣蒸出來的糉子才有透明度及美感。

一般市售大小的糉子約蒸1小時。若做得比較大時，便要多蒸半小時。

If the dumplings are tied too tightly, the glutinous rice will go through to the bean paste. The glutinous rice round the filling will not cook. Wrapping with wafer paper is a precaution step. Bean paste can be frozen before wrapping.

Wrap loosely and the rice should rattle when shaken. This way the cooked dumpling will be transparent and attractive.

The dumplings sold in the market are normally cooked for 1 hour. For larger ones, steam for an extra half hour.

材料	餡料	Ingredients	Filling
糯米2斤（1280克）	鹹蛋黃10個（開邊）	2 catties (1280g) glutinous rice	10 salted egg yolks (split into halves)
鹼水2湯匙	蓮蓉1斤（640克）	2 tbsp lye, 6 tbsp oil	1 catty (640g) lotus seed paste
生油6湯匙	威化紙適量	12 taels (480g) bamboo leaves	Suitable amount of wafer paper
糭葉12兩（480克）		Suitable quantities of reed strips / straws	
水草適量			

1　2　3

4　5　6

製法

1　糭葉、水草用滾水煮15分鐘，取出洗淨瀝乾。

2　糯米洗淨，用清水浸過夜，瀝去水分，加入鹼水及生油拌勻。

3　鹹蛋黃用適量蓮蓉包裹，再以少許威化紙包好，備用。

4　用2塊糭葉交叉折成凹狀，倒入少許糯米，把蓮蓉稍為壓扁，放在糯米上，然後再加入適量糯米，蓋上蓮蓉。

5　再加1塊糭葉於上面封口，用水草鬆動地紮糭，搖動糭子，以能聽到米粒之聲為合。

6　將糭子放入大滾水中（水要浸過糭子面），以大火煲1小時，熄火後，將糭子留在煲內，浸至水凍才將糭子取出。

Method

1　Boil dumpling leaves and reed strips in water for 15 minutes. Remove, wash and drain.

2　Wash glutinous rice and soak in water overnight. Drain and add lye, borax and oil, mix well.

3　Wrap salted egg yolks with suitable amount of lotus seed paste. Cover with wafer paper and set aside.

4　Criss-cross two pieces of bamboo leaves, twist and make a well. Pour in some glutinous rice and add some flattened lotus seed paste on top. Top with more glutinous rice to cover lotus seed paste.

5　Place another piece of bamboo leaf on top to seal. Tie with reed strips loosely. Shake dumpling and you should be able to hear the rice rattling inside.

6　Place dumplings in rapid boiling water (the water should cover all the dumplings). Boil on high heat for 1 hour. Turn off heat and leave dumplings in water. Remove when the water is cold.

Q版五香鹹肉糭
Mini Spicy Dumplings

Q版糭子最近流行於中國各地，由於樣子嬌小可愛，做法簡單又省材料，焓的時間又短，而且快熟，故受一般主婦所歡迎。

現在超市已常有迷你五香糭出售。

Tiny dumpling has become a favourite dim-sum amongst many housewives because of its easy processing method, simple ingredients and short poaching time. Moreover, its cute appearance has made it appealing to the people all over China.

It is widely available in the supermarkets.

材料	醃肉料	Ingredients	Marinade
糯米1斤(640克)	鹽1/2茶匙	640g glutinous rice	1/2 tsp salt
開邊綠豆5兩(200克)	醬油1茶匙	200g splitted green peas	1 tsp thick soy sauce
腩肉6兩(240克)	五香粉1/2茶匙	240g belly pork	1/2 tsp five spices powder
水草12條	糖1/2茶匙	12 reed strips	1/2 tsp sugar
糉葉24片	雞粉1/2茶匙	24 pieces bamboo leaves	1/2 tsp chicken powder
	紹酒1茶匙		1 tsp Shaoxing wine

糯米綠豆調味

鹽1湯匙
雞粉2茶匙
五香粉1茶匙
油3湯匙

Seasonings for Rice & Pea

1 tbsp salt
2 tsp chicken powder
1 tsp spicy powder
3 tbsp oil

1   2

3   4   5

製法

1. 糯米、綠豆分別用清水浸過夜，洗淨瀝乾後加調味拌勻。
2. 腩肉洗淨切成24小粒，拌入醃肉料備用。
3. 水草、糉葉用滾水焓約20-30分鐘，取出洗淨瀝乾。
4. 將1片糉葉折曲成窩狀，放入適量糯米綠豆，加上1粒腩肉，再以糯米綠豆蓋着肉粒，然後將兩邊糉葉向內對摺封口，以水草紮實。
5. 包好糉子後放入大鍋滾水中，焓約1/2小時即成，待水凍後取出，便可供食。

Method

1. Soak rice and peas in water separately overnight. Drain and mix them together with seasonings.
2. Wash and cut belly pork in to 24 cubes. Soak in marinade and set aside.
3. Bring water to boil and poach straw and bamboo leaves for 20-30 minutes. Wash and dry.
4. Make a piece of bamboo leaf into cone, add a layer of rice and pea mixture, put a piece of belly pork and then another layer of rice and pea mixture. Fold both sides of leaf and tie it with a straw securely.
5. Place mini dumplings in boiling water and poach for 1/2 hour over high heat. Set aside until water is cool. Take them out and then serve.

皮料

糯米2斤（1280克）

豬油1湯匙

鹽1茶匙

雞粉1茶匙

荷葉5張

餡料

半肥瘦豬肉4兩（160克）

雞腿肉1隻

冬菇1/2兩（20克）

冬筍4兩（160克）

雞腎（胗）2個

鹹蛋黃4個（1開4）

薑2片

蒜茸1茶匙

醃料

雞粉1/2茶匙

蠔油1茶匙

醬油1茶匙，鹽1/4茶匙

糖1 1/4茶匙

生粉1茶匙，麻油少許

胡椒粉少許，水1茶匙

芡料

薑汁酒1茶匙

蠔油2湯匙

醬油1茶匙

雞粉1/2茶匙

糖1/4茶匙

胡椒粉少許，麻油少許

生粉1茶匙，水4湯匙

Wrapping Ingredients

1280g glutinous rice

1 tbsp lard, 1 tsp salt

1 tsp chicken powder

5 lotus leaves

Filling

160g half lean & half fat
pork, 1 chicken thigh meat

20g tael mushroom

160g taels bamboo shoots

2 chicken gizzards

4 salted duck egg yolks
(cut into quarters)

2 slices ginger, 1 tsp
crushed garlic

Marinade

1/2 tsp chicken powder

1 tsp oyster sauce

1 tsp soy sauce

1/4 tsp salt, 1 1/4 tsp sugar

1 tsp bean flour, Dash of
sesame oil

Dash of pepper, 1 tsp water

Thickening Sauce

1 tsp ginger wine

2 tbsp oyster sauce

1 tsp soy sauce

1/2 tsp chicken powder

1/4 tsp sugar, Dash of pepper

Dash of sesame oil

1 tsp bean flour

4 tbsp water

荷葉必須用水煮過，然後過冷洗淨，不能只用熱水浸，否則包製時容易破損。

糯米的浸泡時間充足才能浸透，浸過夜是較理想的方法。

Boil lotus leaves in water and then rinse in cold water. They cannot just be soaked in hot water otherwise they will break during wrapping.

It is an ideal to soak glutinous rice in water overnight for better result.

製法

1. 糯米洗淨浸過夜，隔去水分，加入材料之豬油、鹽、雞粉撈勻，隔水蒸約30分鐘至飯熟。
2. 冬菇浸軟切粒，冬筍切丁粒後飛水。
3. 豬肉、雞肉切小片，雞腎切薄片後加醃料醃15分鐘後泡油盛起。
4. 燒油2湯匙，爆香蒜茸，傾下冬菇、冬筍炒片刻，瓚薑汁酒，再將已泡油之肉料倒下，炒透加芡汁兜勻上碟待凍。
5. 將荷葉用滾水煮片刻洗淨抹乾，每張剪開3份，攤在檯上。
6. 先鋪一層薄糯米飯，再將適量之餡料放糯米飯內，再蓋上另一層飯把荷葉摺入包好，置蒸籠內，隔水蒸15分鐘即成。

Method

1. Wash glutinous rice and soak overnight. Drain and add lard, salt and chicken powder listed in the ingredients and mix well. Steam over water for about 30 minutes until cooked.
2. Soak mushrooms until soft and dice. Cut bamboo shoots into cubes and blanch in boiling water.
3. Cut chicken and pork into small slices. Slice chicken gizzards thinly and leave in marinade for 15 minutes. Blanch in hot oil and drain.
4. Heat 2 tablespoons of oil and fry crushed garlic. Add mushrooms and bamboo shoots. Fry briefly and splash in ginger wine. Add blanched ingredients and mix well. Pour in thickening sauce, mix and dish. Set aside to cool.
5. Boil lotus leaves in water briefly, wash and dry. Cut each leaf into 3 pieces. Spread on table. Dry with towel. Place on plate.
6. First put on a thin layer of glutinous rice, then a suitable quantity of filling. Cover with another layer of rice before wrapping up with lotus leaf. Place in steamer and steam over water for 15 minutes.

高力豆沙香蕉

Deep-fried Egg White Puffs with Red Bean and Banana

材料

蛋白5隻（約150克）

麵粉20克

粟粉40克，篩勻

豆沙40克

香蕉1/2隻（切粒）

Ingredients

5 (about 150g) egg whites

20g flour

40g cornflour, sifted

40g red bean paste

1/2 (diced) banana

製 法

1 將豆沙分成8等份，每份包入1粒香蕉，搓圓備用。
2 蛋白打成厚忌廉狀。
3 加入篩勻之麵粉和粟粉，拌勻成蛋白糊。
4 用雪糕勺，先舀一半蛋白糊，放入一粒豆沙香蕉。
5 再舀滿一勺蛋白糊，放6-7成溫油中，以慢火浸炸至全熟和呈微黃色，盛起瀝油置碟中。
6 灑上糖霜即成（熱食）。

Method

1 Divide the red bean paste into eight portions. Add a dice of banana into each portion of paste and knead it into a ball.
2 Whisk the egg white into a thick cream.
3 Add in the sifted flour and cornflour and blend into an egg white batter.
4 Use an ice-cream ladle to take half scoop of egg white dough, then add a tiny banana red bean ball in it.
5 Fill the ladle with egg white dough. Then place it in a pan of warm oil. Deep fry over low heat until well done in light yellow. Take out the fried puff and drain out excess oil.
6 Dust some icing sugar over it and serve hot.

材料

金桔約375克

甘草6片

砂糖 1/4杯（約70克）

麥芽糖 3/4杯
（約200克）

清水 1/4杯（約75毫升）

Ingredients

375g tangerines

6 slices licorice

1/4 cup (70g) white
sugar

3/4 cup (200g) maltose

1/4 cup (75g) water

製 法

1 金桔用刀按扁去核，放入清水中煮約3分鐘，盛起過冷瀝乾，再用
   毛巾吸乾水分。

2 砂糖、甘草、麥芽糖、清水同置煲中煮滾。

3 放下金桔，以中慢火煮至糖水收乾，盛起。

4 焗盤放牛油紙，排上金桔，放入焗爐中，以慢火焗乾（約30分鐘）取
   出，待凍即成。

Method

1 Smash the tangerines with a side of a knife and then remove their seeds.
   Boil them in water for about 3 minutes. Take them out, refresh and drain
   out excess water. Pat dry with a towel.

2 Put the sugar, licorice root, maltose and water in a pot and bring to a boil.

3 Put in the tangerines and cook in medium heat until the sweet soup has
   reduced.

4 Place a baking paper on a baking tray. Line the tangerines on it. Put the tray
   into the oven and bake over low heat for about 30 minutes to dry them
   out. Take out the tray, leave it to cool and then serve.

材料

馬蹄粉6兩(240克)

蔗汁4兩(160克)

黃砂糖4兩(160克)

白糖3兩(120克)

清水1520毫升

Ingredients

240g water chestnut starch

160g sugar cane juice

160g brown sugar

120g sugar

1520ml water

製法

1 用480毫升清水調勻馬蹄粉,濾清雜質,分成2份。

2 將清水600毫升置煲中,加入黃砂糖,以大火煮溶,即將一盆粉漿拌勻,隨即將1/4之粉漿撞入煮滾之糖水中,邊撞邊攪勻,調成稀稠適中的糊狀,熄火略凍(約20分鐘),再把餘下之粉漿拌勻加入攪透,再以漏斗濾清雜質。

3 將剩餘之440毫升水和白糖加入蔗汁中煮滾,又將另一盆粉漿拌勻,依照上述做法,先撞1/4粉漿攪勻,熄火待略凍,再加入餘下之粉漿拌透,濾去雜質。

4 將糕盤塗油,置於蒸籠中,放在滾水上,注入一勺粉漿(每注入一次必須攪勻),蓋上蓋2-3分鐘,再用另一個湯勺把另一盆粉漿攪勻注入,如是者輪流交替,二色相間,一層一層地鋪疊蒸熟,直至蒸完為止,待凍切件。

Method

1 Mix 480ml water with water chestnut starch, stir well and then filter out the sediment. Divide batter into 2 portions.

2 Add brown sugar to 600ml water in a saucepan and bring it to boil until dissolved. Stir a portion of batter well, add 1/4 portion of this batter to boiling sugar solution and stir into a smooth batter. Turn off the heat and leave it to cool for about 20 minutes. Then pour in remaining batter, stir it well and then filter out the sediment.

3 Place 440ml water, cane juice and sugar and bring it to boil until dissolved. Process the other portion of batter with the same method mentioned in the step above. Stir it well, add 1/4 portion of this batter in boiling solution to stir into smooth batter. Turn off the heat and leave it to cool for about 20 minutes. Then pour in the rest of the batter, blend and filter out the sediment.

4 Place a greased mould in a steamer with boiling water. Pour in 1/4 cup of stirred brown batter and steam over high heat for 2 to 3 minutes, then add 1/4 cup of stirred sugar cane batter and steam again. Repeat this process until both batters are finished. Let cool, cut in pieces to serve.

材料
蛋白4隻
鹽1/4茶匙
白糖1杯
低筋粉3湯匙
奶粉1湯匙
椰子茸8兩（320克）
雲呢拿香油1/4茶匙

Ingredients
4 egg white
1/4 tsp salt
1 cup caster sugar
3 tbsp low protein flour
1 tbsp milk powder
320g coconut flakes
1/4 tsp vanilla flavour essence

製法
1 先將椰子茸放入焗爐焗5分鐘備用。
2 將蛋白打起，慢慢加入鹽、糖，繼續打至企身和倒放不掉下為止。
3 將麵粉、奶粉、香油及焗好的椰子茸加入，輕輕拌勻。
4 然後分成40等份，每份用湯匙做成圓球。
5 排入焗盤，以170℃焗15分鐘。

Method
1 Roast the coconut flakes in an oven for 5 minutes. Set aside.
2 Whip the egg white. Add in salt and sugar slowly. Whip the mixture continuously until it has become so stiff that it does not fall out when placed upside down.
3 Add in plain flour, milk powder, vanilla flavour essence and roasted coconut powder and then stir lightly.
4 Divide the mixture into 40 portions. Make each of them into a ball with a tablespoon.
5 Line them up on a baking tray and then bake them at 170°C for 15 minutes.

豆沙西米角
Bean Paste Sago Triangles

蒸皮料時,要5分鐘之內,快手攪3次。如蒸過時,表面會凹凸不平並結塊。

攪時必須攪透,否則製成品賣相不美。第一次是蒸2分鐘開蓋,快手攪透;另外兩次則每1分鐘攪一次,共4分鐘;最後的1分鐘離火,快手攪勻便可包餡。

When steaming wrapping, they need to be stirred quickly for 3 times during 5 minutes. If it is teamed for too long, the surface will be uneven and lumpy.

Mix very well when stirring. Otherwise the finish product will not be attractive. Steam uncovered for 2 minutes at the first time, Stir quickly and thoroughly. The other two stirrings should be done a minute apart, making a total of 4 minutes. Turn off heat after the last minute, mix quickly and wrap with fillings.

材料

西米1斤（640克）	
糖5兩（200克）	
生油2兩（80克）	

Ingredients

1 catty (640g) sago
5 taels (200g) sugar
2 taels (80g) oil

餡料

豆沙1/2斤（320克）

Filling

1/2 catty (320g) bean paste

製法

1. 將豆沙搓成欖角形成餡料備用。
2. 先將西米浸1小時，沖洗乾淨，瀝乾水分，加入糖、生油攪勻。
3. 置蒸籠蒸5分鐘（注意：在這5分鐘內，要攪3次，每攪一次必需合上蓋）。
4. 蒸好之西米便可包入餡料，如覺黏手，可以少許生油塗在手中。
5. 將已包餡料之西米角，排放塗油之蒸籠內，再蒸8分鐘即成。

Method

1. Knead bean paste into olive shapes to form filling. Set aside.
2. Soak sago for 1 hour and rinse. Drain and add sugar and oil, mix well.
3. Steam in steamer for 5 minutes (Note: during this 5 minutes, stir 3 times and cover after each stirring).
4. Wrap steamed sage round filling. If it feels sticky, grease hand with oil.
5. Lay wrapped sago triangles in a grease steamer and steam for a further 8 minutes until cooked.

高難度
Advance Pratice

發酵

發酵借助膨脹劑如麵肥（老酵母）、鮮酵母、乾酵母、發粉、梳打粉等，經和麵後形成麵糰，置放一旁令麵糰產生氣泡或脹大的過程。

Fermentation a process of making use of raising agents such as old yeast, fresh yeast, dried yeast, baking powder and soda powder to form dough after mixing, and setting aside to let the dough produce gas or swell.

大酥　　　　　　小酥

摺酥

摺酥以兩種不同質感的麵糰，分為水皮和油皮，經摺疊、覆摺、按壓等複雜程序，令產品產生層次。它分為大酥和小酥，前者即把大份量麵糰的摺疊；後者則以生坯表面掃油脂或包入油心，摺疊成形。

Folding pastry Increase the layers of the products by using two kinds of dough of different textures; they are water dough and oil dough. After the complicated procedures of folding, covering and folding, and pressing, layers are formed. There are big pastry and small pastry. The former involves folding of big dough while the latter involves brushing oil on the surface of raw base or wrapping oil filling and folding to form shape.

複雜技法與成熟
Complicated Technique and Cooking Methods

混色／染色

雙色

蒸

蒸將麵點的生坯放入器具如蒸籠、蒸箱和蒸櫃等，利用水蒸氣的加熱作用使其成熟的烹飪方法。

Steaming It is a cooking method of putting raw base of dim sum into utensils such as steamer, steaming chest and steaming cabinet, and making use of steam to heat until done.

雙色把兩種不同顏色的麵糰，疊在一起，以捲、摺、疊等技法，令成品產生有層次的外觀。

Double colouring Put two kinds of dough of different colours together. By techniques such as rolling, folding and stacking, make the products have a layer appearance.

染色／混色把麵糰滲入天然色素或食用人工色素，令純色麵糰變色，增加成品的外觀。

Colouring / colour mixing Add natural pigments or edible artificial pigments to the dough to make the pure colour dough change colour and enhance the appearance.

煮

煮將生坯投入大量沸水或湯汁的鍋中，利用水的傳熱使製品成熟的烹飪方法。

Stewing It is a cooking method of putting raw base into a large amount of boiling water or broth, and making use of the heat conduction of water to make the products done.

炸

炸把生坯投入大量油的鍋中，利用油脂的熱對流作用，使生坯成熟的烹飪方法。

Deep-frying It is a cooking method of putting raw base into a large amount of oil, and making use of the heat convention of oil to make the raw base done.

煎

煎在平底煎鍋內加入少量油脂，依靠熱傳遞進行成熟的烹飪方法。特點是令成品有香、軟、油潤或光亮的特點。

Shallow-frying It is a cooking method of adding a little oil in a pan, and making use of the heat conduction to make the products done. Its characteristics are making the products fragrant, soft, oily and bright.

烙

烙通過金屬傳導熱量使麵點生坯，透過兩面反複接觸鍋底，直接成熟的烹飪方法，分為乾烙、刷油烙、加水烙。

Baking in a pan It is a cooking method of making use of the heat conduction of metal. The two sides of the raw base of dim sum contact the bottom of the pan alternately to be cooked. There are baking in pan without oil or water (dry baking), baking with oil, and baking with water.

烤

烤以明火或電熱作導熱體，前者能把麵點生坯貼在爐壁上，並將它烤成熟；後者通過遠紅外綫輻射，使生坯成熟。

Grilling It makes use of flame or heat by electricity as heat conductor. In the former case, raw base of dim sum can be stuck on the wall of the hearth, and be grilled until done. In the latter case, the raw base is cooked by the infrared radiation.

烘／焗

烘／焗以焗爐內的高溫把生坯加熱成熟的烹飪方法。特點是溫度高，受熱均勻，令成品色澤鮮明，形態優美。

Roasting / baking It is a cooking method of making use of the high temperature in the oven to heat raw base until done. Its characteristics are high temperature, heating evenly, and making the products brightly coloured and have nice appearance.

蝦餃
Shrimp Dumplings

今次介紹的蝦餃，皮料採用了水晶包皮的用料和製法，此皮料的韌度較強而有彈性，除了吃時口感豐富外，包製時亦能有效地做出美觀造型，使初學者容易掌握包餡的技術。

鮮蝦處理後必須置雪櫃足夠時間，才會爽口好吃，否則蝦肉易「梅」。

The shrimp dumplings introduced here use the same ingredients and method as wrapping for crystal buns. This wrapping is more elastic giving nice chew and a more attractive appearance and is easier for the beginner to get familiar with the technique.

Place the prepared shrimp mixture must be placed in refrigerator for a longer period of sufficient time so as to achieve firmly and nice chew.

皮料	調味料 A	Wrapping Ingredients	Seasonings A
澄麵4兩（160克）	鹽1茶匙	4 taels (160g) tang flour	1 tsp salt
生粉4兩（160克）		4 taels (160g) bean flour	
清水8兩（320克）	調味料 B	8 taels (320g) water	Seasonings B
豬油1/2兩（20克）	糖1.5茶匙	1/2 tael (20g) lard	1.5 tsp sugar
	味粉1/4茶匙		1/4 tsp MSG
餡料	熟油2湯匙	Filling	2 tbsp cooked oil
鮮蝦肉8兩（320克）	胡椒粉少許	8 taels (320g) fresh shrimp meat	Dash of pepper
肥肉粒1兩（40克）		1 tael (40g) pork fat cubes	
馬蹄肉粒1兩（40克）		1 tael (40g) diced water chestnuts	

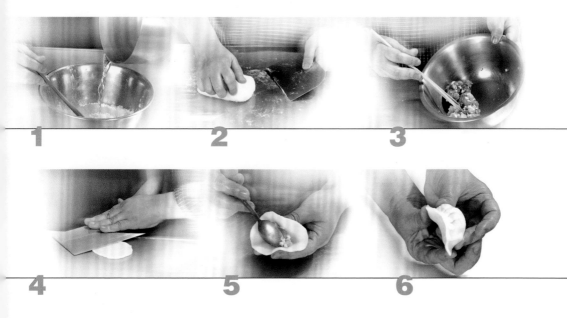

製法

1. 先將澄麵及一半生粉置盤中拌勻，清水煮滾撞落以上粉中，快手以木棍攪勻，使之成為熟粉，覆蓋檯上，5分鐘後加入餘下之生粉及豬油，搓勻成粉糰備用。

2. 鮮蝦肉切粒加入調味 A 料拌勻成膠狀，然後加入肥肉粒、馬蹄粒及調味 B 料拌透成餡料，置雪櫃備用。

3. 粉糰取出搓成長條型，分成小塊，用刀壓成圓型，中心放入適量餡料，包成蝦餃型，放蒸籠內，以大火蒸4分鐘即成。

Method

1. Place tang flour and half the bean flour in a bowl. Boil water and add to flour. Mix quickly with wooden rod to form cooked dough. Cover and after 5 minute add remaining bean flour and lard. Knead into soft dough and set aside.

2. Cut fresh shrimp meat into small cubes and add seasonings A. Mix and stir into paste. Add pork fat cubes, water chestnut cubes and seasonings B. Mix well to form filling. Place in refrigerator.

3. Roll dough into long cylinder shape. Divide into pieces and press into rounds with knife. Place a little filling in the centre and wrap into a shrimp dumpling. Place in steamer and steam on high heat for 4 minutes.

潮州粉果
Chiu Chow Rice Dumplings

貼士 TIPS

必須用大滾水沖入澄麵生粉中，如水沸騰不足，攪勻後的粉會削身。

攪勻後，立即將盆覆轉，蓋於檯上焗5分鐘至透，才在檯上搓至透明同一色調及軟滑。便可分小粒輾皮包餡料。

包時，皮應厚一點成雞冠形。

蒸熟後要塗熟油於面，否則不能久存。

Water must be really boiling when being added to the tang and bean flour. If the water is not boiling, the flour will not bind after mixing.

After mixing, turn bowl upside down immediately and cover on table top for 5 minutes to cook through. Then knead on table top until transparent, evenly coloured, smooth and soft. After that, divide into small balls to make wrapping.

When wrapping, make the outside a bit thicker to form a cock's crown.

Brush cooked oil after cooked. Otherwise they cannot be kept for long.

材料

澄麵5兩（200克）
生粉3兩（120克）
滾水2.5杯（625毫升）
豬油1湯匙

韭菜適量（切度）
乾葱3粒（切粒）

餡料

白肉2兩（80克）
瘦肉6兩（240克）
炸脆花生2兩（80克），壓碎
菜脯2兩（80克），切細粒
蝦米2兩（80克）
沙葛4兩（160克）

調味料

生抽1湯匙
糖1湯匙
鹽1茶匙
蠔油2湯匙
麻油少許
味粉1茶匙
生粉1茶匙
水2湯匙

Ingredients

5 taels (200g) tang flour
3 taels (120g) bean flour
2.5 cups (625ml) boiling water
1 tbsp lard

2 taels (80g) dried shrimps,
4 taels (160g) jicama

some chives (cut into sections), 3 shallots (diced)

Filling

2 taels (80g) white meat
6 taels (240g) lean pork
2 taels (80g) pre-fried peanuts, crushed
2 taels (80g) preserved turnips, cut into small cubes

Seasonings

1 tbsp light soy sauce
1 tbsp sugar
1 tsp salt
2 tbsp oyster sauce
Dash of sesame oil
1 tsp MSG
1 tsp bean flour
2 tbsp water

製法

1 將以上各餡料切細粒，瘦肉用少許生粉及油撈勻。
2 燒油2湯匙，爆香乾葱、菜脯、蝦米加入豬肉炒透，再將餘下之配料加入，韭菜及花生最後落，兜勻打芡上碟待凍。
3 澄麵、生粉篩勻，用2.5杯滾水倒入粉內迅速攪勻，覆蓋檯上，待5分鐘後，倒出置檯上，加入適量油搓成軟滑之粉糰。
4 把粉糰分成小粒，輾成圓形薄皮，放入適量之餡，包成餃子形。放入已搽油之蒸籠內，蒸5分鐘，用熟油塗面即成。

Method

1 Cut the above ingredients into small cubes. Mix pork with a little bean flour and oil.
2 Heat 2 tablespoons of oil and fry shallots, preserved turnips and dried shrimps. Add pork and fry well. Place in remaining ingredients. Lastly add chives and peanuts. Mix and thicken before dishing. Allow to cool.
3 Sieve tang flour and bean flour. Pour in 2.5 cups of boiling water and mix quickly. Wait for 5 minutes and place on table. Add suitable amount of oil and knead to a soft and smooth dough.
4 Divide dough into small balls. Roll into thin round sheets. Place on suitable amount of filling. Wrap into a dumpling shape. Place in greased steamer and steam for 5 minutes. Brush with cooked oil and serve.

萬壽紅龜粿
Long Life Red Turtle Shaped Glutinous Bun

貼士 | TIPS

即客家甜茶粿，龜形模具為取其好意頭。若沒有模具，可做圓型，以葉托底蒸10分鐘即熟。

有人喜歡用蘋婆葉（即鳳眼果葉）托底來蒸，因此葉有獨特的清香。由於不是時常有賣，故多以糉葉或蕉葉取代。

This is a Hakka sweet tea dumpling bun and its turtle shape symbolize lucky. If turtle shaped mould is not available, use normal round bun shape. Place a piece of leave at the bottom and steam for 10 minutes until cooked.

Some people like placing the leaf under the bottom of dumpling because of its solely aroma. However it is seldom found in the market, in such case, you can use the bamboo leaves or banana leaves instead.

128

材料	餡料	Ingredients	Filling
糯米粉5兩（200克）	炒香花生4兩（160克）	5 taels (200g) glutinous rice flour	4 taels (160g) peanuts
粘米粉2.5兩（100克）	糖冬瓜1/2斤（320克）	2.5 taels (100g) rice flour	1/2 catty (320g) sugared winter melon
鹽1茶匙	糕粉1.5兩（60克）	1 tsp salt, 2 tbsp oil	1.5 taels (60g) cake flour
油2湯匙	椰茸1兩（40克）	1 pc leaf for wrapping dumplings	1 tael (40g) desiccated coconut
粿葉1塊	生油1.5兩（60克）	Suitable quantity of red food colouring mixed with 7 oz (200ml) water	1.5 taels (60g) oil
紅色素適量同滾水7安士（200毫升）開勻	清水2湯匙		2 tbsp water

餡料製法 Method for Filling

製法 Method

餡料製法

1. 將花生及糖冬瓜放入攪拌機中打成茸。
2. 糕粉、椰茸、生油和水混合冬瓜茸搓勻成餡料。

製法

1. 將兩種米粉與鹽同篩，置盤中加油及撞入紅色滾水，以筷子攪拌。
2. 稍涼即用手搓勻成軟滑之粉糰。
3. 將粉糰分成12等份，搓圓按扁，包入適量餡料收口，放入模形中，壓實後敲出，在底部放上一塊粿葉。
4. 置蒸籠以大火蒸10分鐘即成。

Method for Filling

1. Place peanuts and sugared winter melon in blender and liquidize.
2. Mix together cake flour, desiccated coconut, oil, water and sugared winter melon to form filling.

Method

1. Sieve the two types of rice flour with salt. Place in bowl and add oil and boiled coloured water. Stir with chopsticks.
2. When slightly cooler, knead with hand to form soft dough.
3. Divide dough into 12 equal portions. Roll into balls and flatten. Place in a suitable quantity of filling and seal. Place in mould and press down hard. Turn out of mould and place a piece of leaf on the bottom.
4. Place in a steamer and steam on high heat for 10 minutes until cooked.

貼士 TIPS

必須用煮滾之糖水撞入粉中才好吃，如用凍水則效果欠佳。

芝麻必須留意用手心搓揉壓實，否則炸時會脫掉影響外觀。

當煎堆浮起表示已熟，但此時如見芝麻仍未夠色，應將火加旺而非將時間加長，因煎堆久炸會裂。

The sugar solution must be boiling when being added to the flour to ensure a good texture. If cold solution is used the end result will be unsatisfactory.

When coating with sesame seeds, press light with palm of hand to ensure the seeds are firmly attached. Otherwise they will fall off during frying.

When the dumplings float to the top, it means that they are cooked. If the colour of the sesame seeds is too pale, turn heat up rather than prolong the cooking time. If the dumplings are overcooked, they will split.

材料	餡料	Ingredients	Filling
糯米粉7兩（280克）	芋茸4兩（160克）	7 taels (280g) glutinous rice flour	4 taels (160g) mashed taro
清水3/4杯（187.5毫升）	砂糖4兩（160克）	3/4 cup (187.5ml) water	4 taels (160g) sugar
白糖3兩（120克）	豬油1湯匙	3 taels (120g) white sugar	1 tbsp lard
芝麻1杯	桂花糖1湯匙	1 cup sesame seeds	1 tbsp osmanehus sugar
	清水2湯匙		2 tbsp water

餡料製法 Method for Filling

煎堆製法 Method for Dumpings

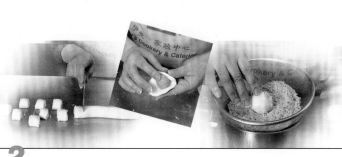

餡料製法

1. 將餡料放鍋中，以中慢火，用鏟不停炒動（約15分鐘）便成芋泥。
2. 取出待凍，即可搓圓作餡用。

煎堆製法

1. 芝麻用清水略沖瀝乾。
2. 水與糖煮滾後加入糯米粉中搓勻成長圓柱形，切成等份，用手搓成圓球按扁，包入1粒芋餡，收口搓圓，然後滾滿芝麻。
3. 以大半鑊油，用中火燒滾，停火片刻，將圓球放入炸至浮起，再轉慢火續炸至金黃即可撈起。

Method for Filling

1. lace filling ingredients in pan and cook on low heat, stirring constantly with spatula (about 15 minutes) to get taro paste.
2. Dish and cool. Form balls to make filling.

Method for Dumpings

1. Wash sesame seeds briefly and drain.
2. Boil sugar and water and add to glutinous rice flour. Mix well and form into long rod. Cut into equal parts. Roll into balls and press flat. Wrap a piece of taro paste in each and seal, Roll in sesame seeds to coat.
3. Using medium heat, bring 2/3 pan of oil to hot and switch off briefly. Place round balls in oil and cook until they float. Change to low heat and continue frying until they are golden brown. Dish.

貼士 TIPS

不愛素食者，可於餡料中加入瘦肉或蝦肉等。

糯米粉加澄麵同搓，只要搓時持久一點，便能搓出韌度，增加口感。

鹹水角用的糖分頗重。但若減去糖分，炸的時間便需要很長，才能炸至深色。

吃不完時，不用浪費，可存放起來，吃時重炸，效果同樣理想及可口。

For non-vegetarians, add pork or shrimp to the filling.

Knead glutinous rice flour and rice flour for a long period of time to create the gluten which enhances the texture.

There is a high content of sugar in the glutinous rice dumplings. If deducting the quantity the sugar, frying time must be lengthen so as to achieve the right dark colour.

Do not waste any leftovers. Store and re-fry before serving. The end result will still be good.

材料

糯米粉4兩(160克)	甘筍1.5兩(60克)
澄麵1兩(40克)	蒜茸1粒
糖1.5兩(60克)	
豬油1兩(40克)	**調味料**
滾水1/4杯(65毫升)	雞粉1茶匙
凍水1/2杯(125毫升)	鹽1/2茶匙
	糖1茶匙
餡料	胡椒粉少許
韭菜1/2兩(20克)	麻油少許
蝦米1/2兩(20克)	生粉1.5茶匙
冬菇1/2兩(20克)	清水2湯匙
沙葛2兩(80克)	

Ingredients

4 taels (160g) glutinous rice flour	80g jicama
	60g carrots
1 tael (40g) tang flour	1 clove crushed garlic
1.5 taels (60g) sugar	
1 tael (40g) lard	**Seasonings**
1/4 cup (65ml) boiling water	1 tsp chicken powder
1/2 cup (125ml) cold water	1/2 tsp salt
	1 tsp sugar
Filling	Dash of pepper
20g chives	Dash of sesame oil
20g dried shrimps	1.5 tsp bean flour
20g mushrooms	2 tbsp water

餡料製法 Method for Filling

製法 Method

餡料製法

1. 將以上所有材料切成細粒及指甲片。
2. 蒜茸起鑊,爆香冬菇及各料,加入調味炒勻盛起待凍。

製法

1. 將滾水沖入澄麵中攪勻。
2. 糯米粉置盤中,加入糖、豬油、凍水同搓勻,再與澄麵混合搓透。
3. 分出粉子,包上餡料。
4. 放中火油內,炸成淺金黃色便成。

Method for Filling

1. Cut the above ingredients into small dices or nail shaped pieces.
2. Fry crushed garlic in oil and fry mushroom and the other ingredients. Add seasonings, mix well, dish and set aside.

Method

1. Pour boiling water into tang flour and mix well.
2. Place glutinous rice flour in bowl, add sugar, lard and cold water. Mix well. Add to tang flour mixture and knead well.
3. Divide flour mixture and wrap in filling.
4. Place in medium hot oil and deep fry until pale brown.

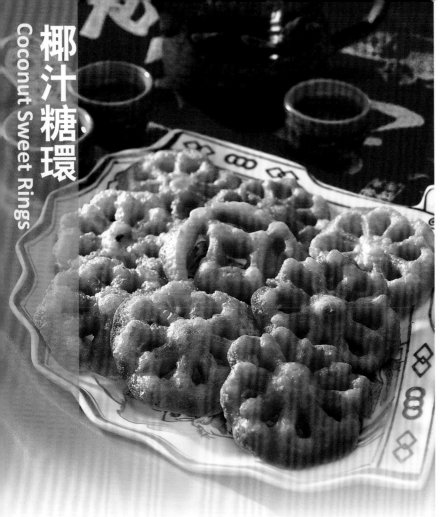

椰汁糖環
Coconut Sweet Rings

材料

麵粉3.5兩（140克）

粟粉3.5兩（140克）
同篩過

雞蛋3隻

糖3.5兩（140克）

椰汁及花奶各1杯，
調勻

Ingredients

3.5 taels (140g) flour

3.5 taels (140g) cornflour
sieved together

3 eggs

3.5 taels (140g) sugar

1 cup each of coconut
milk and evaporated milk,
mixed

製法

1. 蛋糖打起。
2. 將麵粉與椰奶，如梅花間竹般，平分4次加入蛋液中（每加一次，必須打透），直至加完為止，放置1小時後備用。
3. 將糖環模放在油鍋內以中火將油煮至八成滾，隨將模取離油鍋。
4. 輕輕放在粉漿上沾上粉糊。
5. 把黏有粉糊之模隨即轉放熱油中，炸至自動退出。再炸至金黃色即可盛起瀝油，待凍便可放罐中，可保存1-2月。

Method

1. Beat sugar and egg until fluffy.
2. Add flour and mixed coconut milk in batches alternately to egg mixture. This should be done in 4 additions (beat thoroughly after each addition). After all the ingredients are in, set aside for 1 hour.
3. Place ring mould in a pan of oil and cook on medium heat until the oil is medium heat. Remove mould from oil.
4. Coat mould with flour mixture carefully.
5. Place coated mould in hot oil. The cooked rings will separate from the mould after frying. Deep fry until golden brown. Drain well and place in airtight container when cold. They should keep for 1-2 months.

水皮材料
麵粉 4 兩（160 克）
清水 3 安士（100 毫升）

酥心材料
麵粉 4 兩（160 克）
糖 1/4 杯
溶豬油 1 湯匙
南乳 1 湯匙
蒜茸 1/2 湯匙
蘇打粉 1/4 茶匙
清水 1.5 安士（50 毫升）

Ingredients for Water Pastry

4 taels (160g) flour
3 oz (100ml) water

Ingredients for Puff Centres

4 taels (160g) flour
1/4 cup sugar
1 tbsp melted lard
1 tbsp fermented taro curd
1/2 tbsp crush garlic
1/4 tsp baking soda
1.5 oz (50ml) water

製 法

1　將酥心材料之南乳攪勻成糊狀，加入糖、油、蒜茸、蘇打粉、水等攪勻，待糖溶解。
2　加入麵粉調勻，搓成光滑之粉糰。
3　水皮之麵粉加入水後搓成光滑帶些少韌力之麵糰，用毛巾蓋着，醒約 15 分鐘待用。
4　將已醒之麵糰取出輾成薄皮，酥心亦輾至與水皮之長度一樣，但寬度要略比水皮細。
5　掃少許水在水皮上。酥心放在水皮面。
6　向外捲起，置雪櫃，雪至稍硬，切薄片。
7　將切薄之麵片，放滾油中，炸至微黃色，撈起待冷後，便成鬆脆可口之牛耳。

Method

1　Mash red fermented bean curd listed in the puff centre ingredients. Add sugar, oil, crushed garlic, baking soda and water and mix well. Set aside until sugar melts.
2　Add flour and mix well. Knead into smooth dough.
3　Mix flour in the water pastry ingredients with water and knead into a smooth but elastic dough. Cover with tower and set aside for 15 minutes for it to rise.
4　Using the risen dough, roll into a thin pastry. Roll puff pastry to about the same length but a little narrower than the first one.
5　Brush the water pastry with water and place puff pastry on top.
6　Roll outwards and place in refrigerator. When harden, cut into thin slices.
7　Place thinly cut pastry slices in hot oil and fry until light brown. Drain and cool. These will then be crispy cow ear shaped cookies.

篩粉除了去除雜質及清潔外，最重要是使空氣混入，令質感鬆軟。

揉麵糰不能重手大力去搓（這樣成品不會鬆化），應以輕手按揉的方法。

注意，包餡料時勿把餡料黏及角子皮料邊沿，以免炸時散開。

炸角時，由於油溫未及穩定，應先放下2、3隻油角，炸至色澤理想撈起，待第二鍋時，才多放一些。並注意每一鍋放下的油角，必須同時放入及撈出，避免中途加入。

炸好之油角，必須完全放涼，才可用保鮮袋包好入罐，這樣可保存2-3個月。

In addition to getting rid of dirt and keeping clean, it is important to let air mix in flour when sieving, just to make flour soft and puff.

Don't use too much strength when kneading dough, which will not make the finished dough soft. Press and knead in a gentle way.

When wrapping in filling, note not to make filling stick around the edge of pastry, otherwise the pastry will open when frying.

When frying, for the oil temperature is changing, just put in 2-3 triangles first, and then scoop out when they become golden brown. Only in the second batch, can you put in more triangles. Note that the triangles in every batch should be put in and get out at the same time, and do not add new ones when some have been fried for some time.

The fried triangles should be set completely cool before being wrapped in plastic film and contained. Thus it can be stored for 2-3 months.

材料	餡料	Ingredients	Filling
麵粉1斤(640克)	炒香花生肉1/2斤(320克)	1 catty (640g) flour	1/2 catty (320g) fried shelled peanuts
雞蛋2隻(打散)	炒香白芝麻2兩(80克)	2 eggs (beaten)	2 taels (80g) fried white sesame seeds
豬油1杯	白砂糖6兩(240克)	1 cup lard	6 taels (240g) white sugar
清水1/2杯(125毫升)	椰茸2湯匙	1/2 cup (125ml) water	2 tbsp desiccated coconut

1   2   3

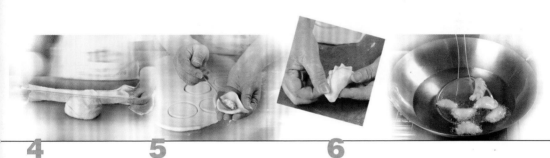

4   5   6

製法

1. 花生研碎，加入芝麻、砂糖及椰茸拌勻成餡料。
2. 麵粉篩在檯上，中間開穴。
3. 加入雞蛋、豬油拌勻，再將清水慢慢加入，邊加邊搓透成柔軟適中的粉糰。
4. 以濕布蓋密，30分鐘後備用。
5. 將粉糰用木棒開薄，用小圓碟扣出圓皮，包入餡料，角邊捏成齒形花邊作收口。
6. 燒油至七成滾，將油角仔投入，以中慢火炸至金黃浮起(約8-10分鐘)即成。

Method

1. Grind peanuts and mix with sesame seeds, sugar and desiccated coconut to form filling.
2. Sieve flour onto tabletop and make a well in the middle.
3. Add eggs and lard and mix. Gradually pour in water, kneading at the same time until a soft dough is formed.
4. Cover with damp cloth and set aside for 30 minutes.
5. Roll dough out with wooden pin and cut rounds with the help of a saucer. Wrap in filling and fold edges in a zigzag pattern.
6. Heat oil to medium hith heat and fry made triangles. Deep fry on low heat until golden brown and the triangles float to the top (about 8-10 minutes).

龍江煎堆
Deep-fried Glutinous Rice Balls

貼士 TIPS

這是我喜愛的食物，小時候見舅母做過，十分懷念，市售的總覺風味不及當年。我喜愛煎堆內的爆穀，它與戲院小食的不同，是沒有殼。以前在雜貨店有售，現在用的人比較少，所以要到九龍城、上環、西環等地的糧油食品店才可買。

炒花生多少可隨意，我在此用的爆穀比花生多，原因是我喜愛吃爆穀。

煮糖膠要注意，可把糖膠滴入凍水中，立即凝固便可，但不要脆身，如軟糖狀即可。

壓圓球餡時切記戴上膠手套，並在手套中心塗上少許油，才不會使餡料黏在手套上。壓時要大力壓實，如壓不夠實，做成的煎堆便不好看。

皮料不要比餡料大太多，否則炸時便會因餡料太少而穿崩。應在皮料上塗少許水，餡料便會黏着皮料，不易散開，此乃製煎堆的竅門。

This is one of my favourite dim-sum. I still missed the ones made by my aunt when I was a kid. I love eating the popcorn inside the glutinous balls. In the old days, it was sold ,without popcorn in it, as a snack in the cinemas. Nowadays, as popcorn has become less popular, it is only available in the grocery stores in Kowloon City, Sheung Wan and Kennedy Town.

The quantity of peanuts used depends on one's own preference. Personally, I prefer using more popcorn.

When making the syrup, you can test its thickness by dripping a droplet in the cold water. If it solidifies immediately into a soft jelly, it is done.

Remember to wear a pair of plastic gloves with some oil brushed on the palm area to avoid sticking the food on it while kneading the peanuts and popcorn balls as filling. Press the ball hard enough or the deep-fried product will not look good.

The glutinous pastry should be only slightly larger than the filling or they will burst out when deep-fried. Remember to brush a little water on the pastry so as to stick the filling firmly in it.

材料A	材料B	Ingredients A	Ingredient B
爆穀(去殼)240克	糯米粉80克	240g popcorn (shelled)	80g glutinous flour
花生肉(炒香)160克	清水1/4杯(62.5毫升)	160g roasted peanuts	1/4 cup (62.5ml) water
芝麻160克		160g seasame	
清水1杯(250毫升)	材料C	1 cup (250ml) water	Ingredient C
片糖360克	糯米粉200克	360g cane sugar	200g glutinous flour
麥芽糖120克		120g maltose	
砂糖80克		80g sugar	
清水2/3杯(167克)		2/3 cup (167g) water	

1 **2** **3** **4**

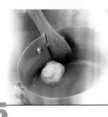  

5 **6** **7** **8**

製法

1 爆穀和花生肉拌勻；芝麻用熱水浸1小時，隔去水分。

2 置清水1杯於鍋中，加入片糖、麥芽糖。

3 以中火煮至濃稠，用筷子沾上糖膠，以拉出絲狀為準，即離火。

4 拌入爆穀及花生肉，快手拌勻，趁熱擠壓成每個約60克重的圓球餡，待用。

5 砂糖用2/3杯水煮溶，把搓成粉糰之B料放入煮熟(約3分鐘)。

6 將材料C置盆中，加入煮熟之粉糰和糖水，快手攪勻搓透，再搓長切粒(每粒約40克重)。

7 每粒再搓揉成圓薄皮，塗上少許水，包入圓球餡料，用力壓實搓成圓形。

8 再塗上適量水分，沾上芝麻並搓實，放中火溫油內，炸至金黃脆皮即成。

Method

1 Toss popcorn and peanut. Soak sesame in hot water for an hour and then drain.

2 Put 1 cup of water in a pot. Add cane sugar and maltose to it.

3 Simmer it under medium heat until sugary mixture turned into a diluted jelly form. Test its thickness by dipping a chopstick to it. Turn off heat if fine lines stuck on chopsticks when pulled out.

4 Add popcorn and peanut to the syrup and mix them by hands quickly. When it is still hot, make some balls of about 60g each to be used as filling.

5 Melt the sugar in 2/3 cup of water. Add the dough from ingredient B and heat it for about 3 minutes to have it cooked.

6 Put the ingredient C in a mixing bowl. Add the cooked dough and sweet soup to it and mix them thoroughly and quickly by hands. Then knead into long dices of about 40g each.

7 Knead the dough dices into small round pastry. On each piece, brush a little water, wrap a piece of peanut and popcorn filling in it and then press into a round shape.

8 Brush some water on it again, sprinkle it with sesame and knead it hard. Deep fry it in lukewarm oil under medium heat until golden brown.

貼士 TIPS

福祿壽三星，寓意吉祥。此糕適合新年及喜慶之用。

開粉漿時必須注意：切忌把粉料置盤中後，將全部糖水1次放入攪拌。這樣會令粉料攪不通透，致使糕的發點偏差，不一定從中間爆開，影響賣相。

應以梅花間竹（1份糖水加1份粉料，約分3次）的方法，每加入1次糖水及粉料必須攪透，當加至最後一次攪透後，應靜置約15分鐘使其發透後再度攪勻，舀入已塗油之模型中，旺火大滾水蒸熟。發糕才會發得完美。

乾果不可多放（可於粉漿舀入杯後，從中間放入數粒），以免過多而使粉漿受壓，造成爆發不理想。

Three Chinese Stars of Longivity, Nobility and Prosperity represent good fortune. This recipe is ideal for the New Year and other happy occasions.

Special attention must be paid when making flour mixture. Do not place flour in bowl and pour in all sugar solution at the same time. This may cause the flour to be saturated at different place and the failure of rising at the centre of the cakes thus affecting the look of the finished product.

Sugar solution and flour mixture should be added in, three separate additions. Every time the ingredients are added, they must be stirred thoroughly before the next addition. After the last mixing, leave it for 15 minutes so that the baking powder can take effect, then stir again. Place in greased mould and steam on high heat over boiling water until cooked. Perfect finishing done.

Another point to notice is not to put in too much dried fruit a few can be put on top of cake mixture before cooking. Too much dried fruit will weigh down the cake while steaming thus affecting the finished look.

材料	Ingredients
麵粉1/2斤(320克)	1/2 catty (320g) flour
糯米粉6湯匙(平匙)	6 tbsp (level) glutinous rice flour
發粉3茶匙(平匙)	3 tsp (level) baking powder
砂糖6兩(240克)	6 taels (240g) sugar
清水1 1/3杯(330毫升)	1 1/3 cups (330ml) water
抹茶粉1湯匙	1 tbsp green tea powder
草莓漿1湯匙	1 tbsp strawberry jam
雜乾果1兩(40克)	1 tael (40g) dried mixed fruit

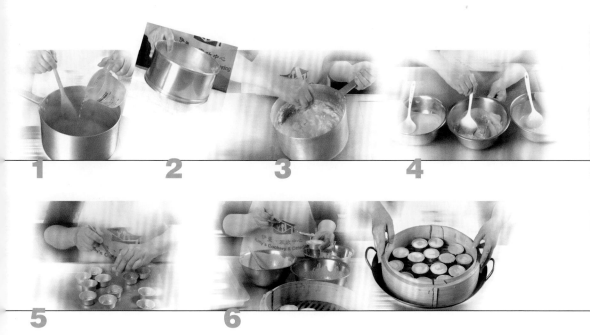

製法

1 先將糖與清水煮滾至糖溶，待冷備用。
2 麵粉、糯米粉及發粉同篩勻。
3 加入已凍之糖水攪勻成稠糊。
4 分成3份，其中2份加入不同味料，再拌入雜乾果。
5 在小碗或錫紙內塗油。
6 把麵糊分放在每一碗內(要放滿為合)，置蒸籠以大火蒸15分鐘，待凍後倒出。

Method

1 Boil sugar in water until melted. Set aside and allow to cool.
2 Sieve together flour, glutinous rice flour and baking powder.
3 Add to cooled sugar solution and mix into thin paste.
4 Divide into 3 equal portions. Mix two portions with the different flavours separately and mixed dried fruit into the third.
5 Grease small bowls or foil containers.
6 Place the different portions of mixtures in separate bowls (to the top of bowls) and steam in steamer over high heat for 15 minutes. Allow to cool and remove from bowl.

貼士｜TIPS

馬拉糕的賣相是糕面鬆發，底部卻較厚。這完全由於材料用上較多的豬油，使其密度較高之故。

黃糖粒子較粗，必須過篩，並篩至足夠的份量為合。

蛋、糖應打至有鬆發感覺，才可分次加入粉料，並注意拌粉漿的方法，絕不能攪，以免生筋使糕發不起來，正確是以捲漿的手法。

撞入之豬油溶液不能熱，暖或凍為佳。撞入後應以輕手快速捲勻，並應注意，油與粉漿必須和勻溶合滲透才可。

蒸糕用之沸水要足夠，中途不可打開蒸蓋加水，否則使糕瀉身。

Steamed sponge cake will have a fluffy top layer and a slightly more solid bottom layer. This is due to the use of lard which increases the density of the mixture.

Brown sugar is comparatively coarse and should be sieved up to appropriate quantity.

Eggs and sugar must be whisked until fluffy before it is added in stages to the flour mixture. This should then be folded in rather than mixing. Mixing can cause the gluten toughened and thus harms the cake from rising.

Make sure there is plenty of water when steaming as water should not be added half way through cooking. If the cover is taken off the cake will not rise.

Make sure there is plenty of water when steaming as water should not be added in half way. If the cover is taken off the cake will not rise.

材料

麵粉 6 兩（240 克）
吉士粉 1.5 兩（60 克）
發粉 2 茶匙
蘇打粉 1/2 茶匙
雞蛋 6 隻
黃砂糖 9 兩（360 克）
雲呢嗱油 1/2 茶匙
豬油 4 兩（160 克）

Ingredients

6 taels (240g) flour
1.5 taels (60g) custard powder
2 tsp baking powder
1/2 tsp baking soda
6 eggs
9 taels (360g) brown sugar
1/2 tsp vanilla essence
4 taels (160g) lard

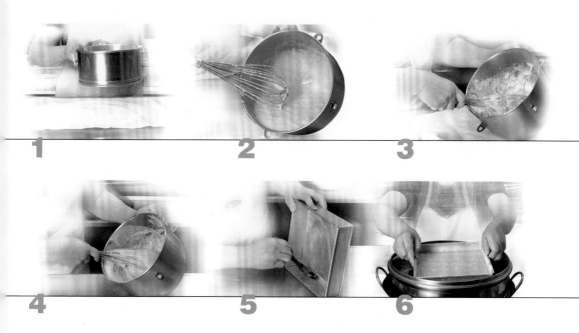

製法

1 麵粉、吉士、發粉及蘇打粉同篩勻 2 次。
2 雞蛋放大盤中，用打蛋器打爛，逐漸加入糖，同打至輕軟成忌廉狀。
3 把已篩勻之粉分 3 次加入蛋液中，以蛋沸輕輕捲勻，放置一旁醒 30 分鐘才可使用。
4 豬油坐溶成流質，加入快手拌勻，再醒 10 分鐘。
5 焗盤塗油撲粉，將粉料倒入，猛火蒸約 50 分鐘即成。

Method

1 Sieve together flour, custard powder, baking powder and baking soda twice.
2 Place eggs in a large bowl and whisk with egg beater. Gradually add brown sugar and beat until creamy and forms soft peaks.
3 Add sieved ingredients in three stages, beating slightly after each addition. Set aside for 30 minutes before using.
4 Sit lard in hot water until melted and mix into the above quickly. Leave for another 10 minutes.
5 Grease and flour baking tin and pour in mixture. Steam on high heat for 50 minutes until cooked.

材料 A

麵粉 336 克
酵母粉 2/3 湯匙
糖 1 茶匙
溫水 1/3 杯（80 毫升）

材料 B

白糖 1/2 杯
豬油 3 湯匙
蛋白 1 個
發粉 2 湯匙
醋 1/2 茶匙
麵粉 112 克

叉燒飽餡料

叉燒 320 克（切粒）
乾葱肉 40 克（切粒）
生油 40 克

芡料

生粉 80 克
清水 160 克

調味

清水 480 克
蠔油 40 克
生抽 40 克
老抽 80 克
糖 200 克
鹽 1 茶匙
味粉 1 茶匙
胡椒粉少許
橙紅粉少許

Ingredients A

336g plain flour
2/3 tbsp instant yeast
1 tsp sugar
1/3 cup (80ml) warm water

Ingredients B

1/2 cup sugar
3 tbsp lard
1 egg white
2 tbsp baking powder
1/2 tsp vinegar
112g plain flour

Filling for Cha Shiu Bun

320g diced Cha Shiu
(Barbecued Pork)

40g diced shallots

40g oil

Gravy

80g cornflour
160g water

Seasonings

480g water
40g oyster sauce
40g light soy sauce
80g dark soy sauce
200g sugar
1 tsp salt
1 tsp msg powder
Pinch of pepper

Pinch of orange and red
food colouring

製法 Method

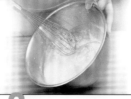

1 **2**

叉燒包餡料製法 Making of Cha Shiu Bun Filling

3 **1** **2**

製法

1. 酵母、糖放溫水中，靜置5-10分鐘，使其發酵，當水面浮上一層泡沫時即可使用。
2. 將發酵水倒入盤中與麵粉一起搓揉成糰，把麵糰放在抹油盤中蓋以白布，讓麵糰發酵為原來體積約兩倍大時即可使用。
3. 將麵糰取出放在檯上，中間開穴，把材料B加入，搓成一大麵糰至光滑均勻後，分切成16小塊，每小塊壓扁，包入餡料，墊上底紙入蒸籠內，大火蒸12分鐘即熟。

叉燒包餡料製法

1. 用1湯匙生油爆香乾葱，加入調味煮滾片刻，撈起乾葱，再以生粉水埋芡，倒出待涼備用。
2. 將叉燒加入芡料中拌勻即成餡料。

Method

1. Soak yeast and sugar in warm water for 5-10 minutes to prove until bubbles come out.
2. Sift flour and pour in yeast mixture and knead into a soft dough. Place it in a greased bowl and cover with a towel. Set it aside to prove in a warm place until it becomes double in size.
3. Place the dough on the table and make a hole in centre. Stir in the mixed ingredient B, knead into smooth dough and divide into 16 pieces. Flatten each piece and wrap 1 portion of cha shiu filling in it. Stick a small square paper at the bottom of each bun, steam over high heat for 12 minutes and serve hot.

Making of Cha Shiu Bun Filling

1. Heat up the wok with 1 tbsp of oil and saute the shallots. Add in mixed seasonings and cook for a while. Remove the shallots. Thicken the sauce with the gravy mixture. Set aside to cool.
2. Add the Cha Shiu in the sauce and mix evenly.

調味芡汁料做好可放冰箱，待用時才加入叉燒，用多少做多少，可節省餡料。

Gravy can be stored in the refrigerator before use. Add the barbecued pork to it when making the buns. To save the ingredients for filling, just use what you need.

點心TIPS

貼士｜TIPS

一般發麵糰如加了依士，通常要發酵1小時才可以做包。但此壽包用了大量發粉，所以不用發酵也可立即做。

色素不要沾得太多，否則滴在包面便不好看，如想顏色自然一點，可以用吸水紙巾先把包子的一半蓋着才彈上顏色，便會有很好的效果。

If yeast is added to a dough, it takes an hour for the dough to prove before making it into buns. To speed up the process, use more baking powder.

Don't splash too much colouring on the buns or it will ruin the appearance. For a more natural look, cover half of the bun with a paper towel before sprinkling the colouring to it.

皮料	餡料	Pastry Ingredients	Filling
麵粉 10 兩（400克）	蓮蓉 1斤（640克）	400g plain flour	640g lotus seed paste
發粉 1.5 茶匙	鹹蛋黃（蒸熟）4隻	1.5 tsp baking powder	4 preserved egg yolks (steamed)
砂糖 1 兩（40克）		40g sugar	
臭粉 1/5 茶匙		1/5 tsp amonia powder	
依士 1 茶匙		1 tsp yeast	
豬油 1 湯匙		1 tbsp lard	
桃紅色素（開水）少許		Crimson red food colouring (mixed with some water)	

1 2 3 4

製法

1 麵粉、依士、發粉同篩勻三次，開穴。
2 加入糖、臭粉，於穴中拌勻，放下豬油及溫水搓成粉糰，發烤1小時備用。
3 鹹蛋黃切成小粒，將蓮蓉分成小塊，中間放入鹹蛋黃成餡料。
4 將粉糰分成小塊，搓成圓形，中間包入餡料搓成壽包形。
5 放入蒸籠蒸約8分鐘，取出。
6 用小刀背壓一坑紋成壽桃形。
7 用牙刷沾上少許色素，彈在壽包面即成。

Method

1 Sift flour and baking powder together for 3 times. Make a well.
2 Put sugar, egg white and white vinegar in the hole and mix. Add in lard and warm water. Knead into a dough.
3 Cut the preserved egg yolk into small dices. Divide the lotus seed paste into small pieces and wrap an egg yolk dice in each as filling.
4 Knead the dough into several small round pastries. Place the filling in the centre and wrap into a peach shape.
5 Put in a steamer and steam for about 8 minutes. Take it out.
6 Cut a slash on the bun with the back of a small knife to make it into the shape of peach.
7 Splash some colouring on the buns with a toothbrush and serve.

迷你臘腸卷
Mini Chinese Sausage Bun

貼士 | TIPS

快速伊士適合與粉料同篩勻。如用普通的伊士，必須用約1/2-1茶匙糖，以溫和暖水開勻，置放5-10分鐘待其發酵才可以用。

麵粉的新與舊、天氣是否潮濕，與面糰有密切關係。北風天，濕度低，粉會較乾，水分要用多點；回南的天氣濕度高，粉便較濕，水分便可減少些。

搓面糰時，準備多一些水量是十分重要的，除了因為方便加減外，亦可避免因中途水量不夠時才去加添，而導致溫度不統一而影響面糰的質素。

搓好的面糰，用濕布蓋於盆中1-1.5小時後，會發酵至兩倍大。可用手指蘸少許乾粉，從面糰中間插入，如成肚臍般凹入而不會彈起復原的，即已成功發酵。

蒸包的時間長短，視乎包皮的厚薄程度。可用手輕按，熟透的包子有回彈能力，反之則未熟透。蒸包要猛火，水要大滾，未夠鐘不得打開蓋，蒸出來的包才夠潔白。

Instant yeast is suitable for sieving with flour. For ordinary yeast, they must be mixed with 1/2-1 teaspoon of sugar and warm water. Leave for 5-10 minutes to become frothy.

The freshness of the flour and moisture in the air play an important part in making the dough. Windy and low humidity winter day, the flour will be drier and requires more water. High humidity spring day, the flour is more moist and less water should be used.

When kneading dough, it is better to prepare more water than required, for which bringing convenience and keeping consistent water temperature. Otherwise, input of water in different temperature for kneading will affect the quality of the dough.

The kneaded dough will double in size after being left in a bowl covered with wet cloth for 1-1.5 hours. Use finger dusted with a little flour and poke into middle of dough. If the dent does not bounce back the dough has finished fermenting. It can then be divided into portions to make buns. Make into balls and leave for 15-20 minutes before rolling.

The steaming time varies according to the thickness of the wrapping. Press lightly with hand with re-bouncing means buns are cooked. The opposite means they are not cooked. Steam on high heat with rapid boiling water. Do not remove lid in half way. The cooked buns will then be white in colour.

材料	餡料	Ingredients	Filling
麵粉1磅(480克)	臘腸10條	1 lb (480g) flour	10 pcs Chinese sausages
快速依士1湯匙		1 tbsp instant yeast	
糖1/2杯		1/2 cup sugar	
豬油3湯匙		3 tbsp lard	
蛋白1隻		1 egg white	
白醋1/2茶匙		1/2 tsp white vinegar	
溫水7安士(200毫升)		7 oz (200ml) warm water	

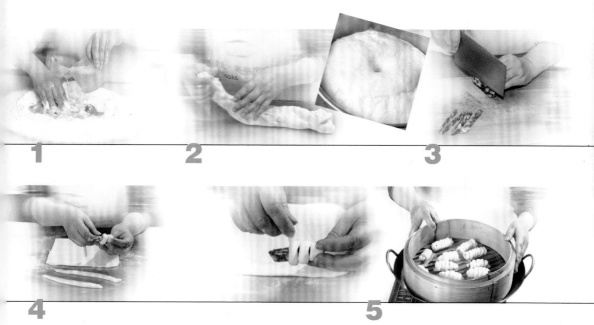

製法

1 麵粉、依士同篩勻，中間開穴，放入糖、醋、蛋白，慢慢加入適量之溫水及豬油，搓成光滑均勻的軟糰，放桌上搓至有彈性。

2 搓好的麵糰放在抹油的盆內，覆蓋一塊白布，放溫暖地方發1小時。

3 臘腸洗淨一切開8條。

4 取出麵糰成0.8公分厚的長方塊，用利刀切成1公分寬的長形麵條狀。

5 將臘腸放在長形麵條的一頭，然後捲起成螺絲形狀，以紙墊底，排放在蒸籠內，猛火蒸12分鐘。

Method

1 Sieve flour and yeast. Make well in centre and add sugar, vinegar and egg white. Gradually add suitable quantities of warm water and lard. Knead to form smooth, even and soft dough. Throw dough on table repeatedly until elastic.

2 Place dough in a greased bowl. Cover with white cloth and leave in a warm place to rise for 1 hour.

3 Wash Chinese sausages and cut into 8 pieces each.

4 Take dough and make into 0.8 cm. thick rectangles. Using a sharp knife, cut into 1 cm. wide strips.

5 Place Chinese sausage at one end of flour strip and roll into a snail shape. Line bottom with paper and place in steamer and steam on high heat for 12 minutes.

菠蘿在世界各地均有出產，尤以中國廣東和台灣的產量最豐富。由於菠蘿的末端有一撮像羽毛般的細葉，與鳳尾甚為相似，台灣人便稱它為「鳳梨」。

由於此酥的餡料是生果製成，所以熱食較為酸，不甚可口，最好放涼後才吃，便會柔韌好吃有質感。

此處教大家自製餡料，雖然較為繁瑣，但十分實用且有滿足感。如在台灣，餡料是有現貨出售不用自製。

Amongst the various pineapple growers all over the world, Canton of China and Taiwan are the largest producers of this fruit. As there is a bunch of feather-like small leaves on its end, like a tail of phoenix, it is called hoenix pear in Taiwan.

Since the filling of this pastry is made of fruit, it tastes a bit sour when it is hot. It'd better to have it cooled down to enable a chewy texture before serving.

Despite the tedious process of this homemade pastry, the method introduced here is very practical and you will have a sense of satisfaction when it is made. In Taiwan, ready-made pastry filling is available.

材料	餡料	Ingredients	Filling
鹹蛋4隻	糖水菠蘿(隔去糖水)12片	4 salted eggs	12 slices pineapple in syrup (drained)
麵粉約340克	糖冬瓜約160克	340g plain flour	160g preserved winter melon
牛油約230克	糖30克	230g butter	
糖霜約115克	油1湯匙	115g icing sugar	30g sugar
	糕粉約80克		1 tbsp oil
塗面用料	菠蘿味香油1/4茶匙	**Coating**	80g cooked glutinous rice flour
蛋黃1隻		1 egg yolk	1/4 tsp pineapple flavour essence
清水1茶匙		1 tsp water	

1 2 3 4 5

餡料製法

1. 先將菠蘿放進攪拌機中，打爛後壓出液汁。取回1/2杯液汁及160克菠蘿渣留用。
2. 將糖冬瓜、糖、菠蘿汁及香油等置攪拌機中打成漿。
3. 將以上打好之材料放入盤中，加入菠蘿渣、油及糕粉攪勻便成餡料。

製法

1. 鹹蛋蒸熟，去蛋白，將蛋黃壓爛成茸。
2. 麵粉篩入盤中，與牛油、糖霜、蛋黃等拌勻，搓成軟糰，鬆身15分鐘後再置冰箱，雪約40分鐘備用。
3. 將麵糰搓成長條，分成27等份，每份重量約為25克。
4. 餡料重量每份約15克，餘下餡料可放冰箱，貯存期約2星期。
5. 將麵糰按薄包入餡料，放入已灑粉之餅模中按平，將餅模左右輕敲，再向中央位置一敲倒出，排放在焗盆上，放入預熱的焗爐中，以190℃火力焗約15-20分鐘，或直至酥身呈金黃色即成。

Making Of Filling

1. Place the pineapple in a blender, mash it into a paste. Set aside 1/2 cup of pineapple juice and 160g pineapple puree.
2. Place preserved winter melon, sugar, pineapple juice and oil in a blender and mash into paste.
3. Pour the paste in a mixing bowl and stir in pineapple puree, pineapple flavour essence and cooked glutinous rice flour to form the filling.

Method

1. Steam salted eggs till cooked, remove egg whites, the egg yolks mash into paste.
2. Sift flour into a mixing bowl. Toss with butter, icing sugar and egg yolk paste. Knead into a dough. Set aside for 15 minutes and then chill in refrigerator for 40 minutes.
3. Roll the dough into a long cylindrical stick. Divide it into 27 portions of about 25g each.
4. Divide filling into portions of 15g each. Keep the remaining in refrigerator. The filling can be stored up to 2 weeks.
5. Flatten the dough into a pastry and wrap in filling. Stuff in a mould dusted with flour. Knock lightly to drop it out and arrange on a baking tray. Bake in a preheated oven at 190°C for 15-20 minutes until golden brown.

貼士 TIPS

餡料要雪的原因是使其收乾水分，容易包入做型。

餡料加入少許水同攪至起膠，肉便吸收了水分；當
包子蒸好時，水分會從肉中釋出，吃時便會有濕潤
口感。

包子收口部分要壓在盆下，使汁不能溢出。

煎包宜用易潔鑊，排列不宜太密，因熟後體積會大
一點。反面後不用再加水，因包已熟，只待煎至金
黃即可供食。

The reason for chilling the filling is to make it dry and
easy to be wrapped.

A little water is added to the meat filling to make it sticky.
While steaming the buns, filling will water and the buns
will be moist when eaten.

The seam of the buns should face down so that the meat
juice will not leak out.

It is better to use a non-stick pan. Do not crowd the
buns, as they will expand when cooked. After turning,
there is no need to add water as the buns have already
cooked, cook until turning golden brown before serving.

材料
筋粉4兩(160克)
麵粉4兩(160克)
依士1茶匙
糖1湯匙
鹽1/2茶匙
溫水132毫升
豬油1湯匙

餡料
牛肉1/2斤(320克)攪碎
葱花2兩(80克)

調味料
糖1茶匙
鹽1/2茶匙
醬油1湯匙
胡椒粉1/4茶匙
蛋白1隻
麻油1茶匙
酒1茶匙
味粉1茶匙
生粉1/2茶匙

Ingredients
4 taels (160g) bread flour
4 taels (160g) flour
1 tsp Yeast
1 tbsp sugar
1/2 tsp salt
132ml. warm water
1 tbsp lard

Filling
1/2 catty (320g) minced beef

2 taels (80g) chopped
spring onions

Seasonings
1 tsp sugar
1/2 tsp salt
1 tbsp soy sauce
1/4 tsp pepper
1 egg white
1 tsp sesame oil
1 tsp wine
1 tsp MSG
1/2 tsp bean flour

1   2   3

4   5   6

製法
1. 筋粉、麵粉、依士同篩在桌上，開穴。
2. 放上糖、鹽在穴中，注入適量溫水及豬油，搓成光滑麵糰，放在抹油盤中，30分鐘後備用。
3. 將發酵好之麵糰分割成12-14等份，滾圓後再發酵15分鐘。
4. 餡料預先加入調味及葱花攪透至起膠置雪櫃備用。
5. 將麵糰兩面黏上麵粉，用麵棍由邊緣向內輾成邊薄中間較厚之麵皮，包入適量 肉餡（約25-30克），注意接合處需捏緊。
6. 平底鍋中放入適量生油，以慢火煎片刻，注入2湯匙清水合蓋8分鐘，翻轉壓扁再煎另一面至金黃即可，合計所需時間為10-15分鐘。

Method
1. Sieve bread flour, flour and yeast on table and make well.
2. Add sugar and salt into well. Add suitable amount of water and lard. Knead into smooth dough. Place in greased bowl and set aside for 30 minutes.
3. Divide risen dough into 12-14 equal portions. Roll into balls and allow to rise for another 15 minutes.
4. Mix filling with seasonings and chopped spring onions. Stir until sticky and place in refrigerator.
5. Coat both sides of dough with flour. Roll with pin from outside in to for a sheet that id thin on the outside and thick in the middle. Wrap in suitable amount of meat filling (about 25-30g) and remember to squeeze the opening tightly together.
6. Place required amount of oil in a flat bottom pan. Fry on low heat for a while and add 2 tablespoons of water. Cover for 8 minutes. Turn and press flat. Fry the other side until golden brown. The total cooking time is 10-15 minutes.

椰子花卷
Coconut Flower Bun

材料

筋粉1磅（480克）

快速依士1湯匙

加強劑3/4茶匙

糖3安士

奶水2湯匙

蛋1/2隻

牛油5湯匙（約70克）

溫水1杯（250毫升）

椰茸1杯，砂糖1/2杯，混合使用

蛋黃2隻，糖水2湯匙，塗面用

蛋3隻，打成蛋液

Ingredients

1 lb (480g) bread flour

1 tbsp instant yeast

3/4 tsp flour improve

3 oz sugar

2 tbsp milk

1/2 egg

5 tbsp (about 70g) butter

1 cup (250ml) warm water

1 cup desiccated coconut, 1/2 cup sugar, mix to use

2 egg yolks, 2 tbsp syrup, for glazing

3 eggs, beaten

製法

1　將所有材料混合一起，搓揉成軟滑麵糰，發酵1小時或至兩倍大，分切成20等份，滾圓後續發酵20分鐘擀薄至15毫米，待鬆根3分鐘。

2　麵糰對折，切割4-5刀，成條狀，連續可切至數個後，再整形。

3　再將切好鬆弛的麵糰輕輕「拉長」放入全蛋液內，使整個麵糰黏滿蛋液。

4　將黏滿蛋液的麵糰，直接放入加糖的椰子粉內，兩面同時黏滿椰子粉。

5　再把麵糰輕輕「拉長」，同時以反方向動作，將麵糰捲起，不宜過緊。

6　將捲起的麵糰打成結，打結時不宜過緊，以保持鬆狀為宜。

7　打好結之麵糰，平均放置焗盤中，進行最後發酵。

8　待麵糰脹大至2-3倍量時，放進焗爐，以230℃焗8分鐘，取出掃蛋黃糖水。

9　如焗爐有上下火度數可用，上火230℃、下火200℃烤焗。

Method

1　Combining all the ingredients together and knead it as a dough. Wait for an hour for the dough to enlarge to its double Then, divide the dough into 20 small portions and roll them into small balls. Wait for 20 minutes, slide them into 15mm then, and give them another 3 mintues more.

2　Fold relaxed dough pieces together and cut 4-5 slashes with bread knife into strips. A few pieces can be cut before assembling.

3　Stretch relaxed and cut strips gently and dip in egg until fully covered.

4　Place egg coated strips into sugar and coconut mixture. Cover both sides with mixture.

5　Stretch strips gently again and roll in an anticlockwise direction. Do not roll too tightly.

6　Tie rolled strips into knots. Knots should be loose rather than tight.

7　Place knotted flour dough flatly on tray for final rising.

8　When they have expanded to 2-3 times their original volume, place in oven and bake for 8 minutes at 230°C. Remove and brush with egg and sugar solution.

9　If the oven has different top and bottom settings, have 230°C at the top and 200°C at the bottom.

材料

牛油 8 安士（240 克）

雞蛋 5 兩（200 克）

砂糖 6 兩（240 克）

核桃肉 1/2 杯

麵粉 8 兩（320 克）

鹽 1/8 茶匙

雲呢嗱香油 1/2 茶匙

蛋黃色素 1/4 茶匙

Ingredients

8 oz (240g) butter

200g eggs

240g sugar

1/2 cup shelled walnuts

320g flour

1/8 tsp salt

1/2 tsp vanilla essence

1/4 tsp yellow food colouring

製法

1 牛油攪成奶油狀，慢慢加入糖，繼續攪至奶白色。

2 雞蛋分 3 次加入牛油內攪透。

3 麵粉、鹽同篩勻 3 次。

4 留 1/4 杯麵粉與核桃肉撈勻。

5 將篩好之麵粉分 3 次加入牛油內，輕輕捲勻，最後加入已拌入麵粉的核桃肉，捲透。

6 用油塗焗盆，撲上麵粉，將成分倒入，呈兩邊高中間低狀。

7 放進已預熱焗爐 200℃ 焗 30 分鐘。

Method

1 Cream butter and gradually add sugar. Keep stirring until pale white in colour.

2 Add eggs in three separate batches and mix well with butter.

3 Sieve flour and salt three times.

4 Save 1/4 cup of flour and mix with shelled walnuts.

5 Add sieved flour to butter in three batches and fold gently. Lastly put in flour coated walnuts and fold in well.

6 Grease baking tin with oil and coat with flour. Place ingredients in tin and hollow out centre with the sides higher.

7 Place in pre-heated 200°C oven and bake for 30 minutes.

蒸雞蛋糕
Steamed Sponge Cake

材料

自發粉140克
吉士粉15克
雞蛋5隻
糖200克
鹽1/4茶匙
粟米油45克

Ingredients

140g self-raising flour
15g custard powder
5 eggs
200g sugar
1/4 tsp salt
45g corn oil

製法

1. 將自發粉、吉士粉同篩勻。
2. 雞蛋打勻,加入糖、鹽打成厚忌廉狀。
3. 將粉分三次加入,輕手拌勻,再加入粟米油拌透。
4. 糕盤塗油,撲上麵粉,將材料倒入,立即放入蒸籠內,以大火蒸約30分鐘。

Method

1. Sift self-raising flour and custard powder together.
2. Beat the eggs. Add in sugar and salt and whip up a thick cream.
3. Divide the flour into 3 portions and flod them in the batter one after the other. Mix it lightly by hands. Add in corn oil and blend.
4. Brush some oil on a cake mould, sprinkle some flour on it and then pour in the batter. Put it in a steamer and steam it over high heat for about 30 minutes.

材料

麵粉 1 斤（640克）

發粉 1 兩（40克）

吉士粉 1.5 兩（60克）

椰漿 2.5 兩（100克）

花奶 1/4 杯（62.5毫升）

糖 14 兩（560克）

雞蛋 20 兩（800克）連殼計

雲呢嗱粉 1 安士（30克）

雲呢嗱香油少許

豬油 10 兩（400克）坐溶

餡料

肉鬆適量

Ingredients

1 catty (640g) flour

1 tael (40g) baking powder

1.5 taels (60g) custard powder

2.5 taels (100g) coconut milk

1/4 cup (62.5ml) evaporated milk

14 taels (560g) sugar

20 taels (800g) eggs including shells

1 oz (30g) vanilla powder

Dash of vanilla essence

10 taels (400g) lard, melted

Filling

Suitable quantity of pork floss

肉鬆鳳凰卷
Pork Floss Egg Rolls

製法

1 將麵粉、發粉及吉士粉同篩勻。
2 雞蛋打在盤中，加糖同打至糖溶。
3 將麵粉料分次加入蛋糖中，邊加邊打透。
4 將椰漿、花奶、雲呢嗱香油加入打透之麵糊中。
5 最後落豬油。以手充分拌透即成。
6 將蛋卷專用之圓鐵板燒熱，舀入1茶匙麵糊，合上鐵板約10秒打開。
7 放上適量餡料，做成小長方形枕包狀即成。

Method

1 Sieve together flour, baking powder and custard powder.
2 Place eggs in bowl and beat with sugar until melted.
3 Add flour to egg and sugar in batches, beating all the time.
4 Add coconut milk, evaporated milk and vanilla essence to well beaten flour mixture.
5 Lastly add lard and mix well with hands.
6 Heat an egg roll hot plate and place 1 teaspoon of egg mixture on top. Close plates and open after 10 seconds.
7 Put in suitable amount of filling and roll into a small pillow shape.

157

鬆化蛋散

材料
麵粉 150 克
高筋粉 38 克
雞蛋 2 隻
臭粉 1/3 茶匙
蘇打粉 1/4 茶匙

糖膠料
砂糖 225 克
麥芽糖 75 克
醋 1.5 茶匙
清水 3/4 杯

Ingredients
150g plain flour
38g high protein flour
2 eggs
1/3 tsp ammonia powder (edible)
1/4 tsp soda powder

Syrup Ingredients
225g sugar
75g maltose
1.5 tsp vinegar
3/4 cup water

製法
1. 麵粉、根粉同篩在檯上，開穴。
2. 將臭粉、蘇打粉、雞蛋放穴中拌勻，再將四周的粉撥入，輕手搓成光滑之麵糰，以濕布蓋着，醒（即使麵糰鬆弛）30分鐘。
3. 麵糰輾薄，摺成三幅，再輾薄，重複兩次。
4. 把麵糰輾成薄皮，切成長方形，兩條麵皮對疊，在中央剗一刀。
5. 將一端從中間之刀痕穿入，在底部拉出，放入八成滾油中炸至浮起及兩面微黃撈起，瀝油。

糖膠製法
將糖膠料同置煲中，慢火煮至可以拉出絲，即可淋在蛋散上。

Method
1. Sift both plain flour and high protein flour together on the table. Make a hole in its centre.
2. Put ammonia powder, soda powder and eggs into the hole and mix together with the surrounding flour. Knead the mixture lightly to make a smooth dough. Cover it with a wet cloth and leave it for 30 minutes.
3. Roll the dough into a sheet. Fold it into 3 layers and then roll it again. Repeat twice.
4. Roll the dough into a thin sheet. Cut it into rectanglar pieces. Fold the sheet into 2 equal halves and slit in the middle.
5. Slide one end of the sheet through the slit. Put it into a wok of heated oil and deep-fry until light brown on both sides. Remove and drain out excess oil before serving.

Making of syrup
Put the syrup ingredients in a pot. Simmer it over low heat until thin silky lines can be pulled out from the syrup. Pour it over the egg crisps and serve.

材料 A
麵粉 120 克
粟粉 2 湯匙
吉士粉 4 湯匙
清水 1 杯

材料 B
雞蛋 5 隻
花奶 3 湯匙

材料 C
牛油 120 克
糖 320 克
清水 1 2/3 杯（410 毫升）
煉奶 2 湯匙

Ingredients A
120g plain flour
2 tbsp cornflour
4 tbsp custard powder
1 cup water

Ingredients B
5 eggs
3 tbsp evaporated milk

Ingredients C
120g butter
320g sugar
1 2/3 cup (410ml) water
2 tbsp condensed milk

製 法
1 將材料 A 之粉料同篩勻，置盆中，加入清水攪透。
2 將材料 B 之雞蛋、花奶同打起，加入攪透之材料 A 成混合料。
3 將材料 C 之牛油、糖及清水放煲中煮沸，加入煉奶攪勻。
4 將混合料倒下，快手攪成蛋麵糊。
5 焗盤塗油，墊上牛油紙，把蛋麵糊倒入抹平。
6 放在注有水之焗盤中，以 180℃ 中火焗 40-50 分鐘，凍後切件。

Method
1 Sift the plain flour, cornflour and custard powder of Ingredients A together. Put it in a mixing bowl, add in water and stir thoroughly.
2 Whip the eggs and evaporated milk of Ingredients B. Add in the mixture from A and stir thoroughly.
3 Melt the butter, sugar and water of Ingredients C in a pot and boil. Add in condensed milk and stir thoroughly.
4 Mix all the ingredients together and stir it vigorously into a batter.
5 Brush a baking container with oil, line it with baking paper. Pour in the batter and level its surface.
6 Place it on a baking tray filled with water. Bake it in an oven at 180°C for 40 to 50 minutes. Cut it into pieces when it is cool.

圖解點心製作教程

An Illustrated Guide to Dim Sum

編著	Author
蔡潔儀	Kitty Choi

編輯	Editor
郭麗眉	Cecilia Kwok

翻譯	Translator
翠詩 Clara	Tracy Ip Clara Ho

攝影	Photographer
幸浩生	Johnny Han

封面設計	Cover Designer
梁施敏	Mandi Leung

版面設計	Designer
劉紅萍	Pancy

出版者　Publisher
萬里機構・飲食天地出版社　Food Paradise Publishing Co., an imprint of Wan Li Book Company Limited
香港鰂魚涌英皇道1065號東達中心1305室　Room 1305, Eastern Centre, 1065 King's Road, Quarry Bay, Hong Kong.
電話　Tel: 2564 7511
傳真　Fax: 2565 5539
網址　Web Site: http://www.wanlibk.com

發行者　Distributor
香港聯合書刊物流有限公司　SUP Publishing Logistics (HK) Ltd.
香港新界大埔汀麗路36號中華商務印刷大廈3字樓　3/F, C & C Building, 36 Ting Lai Road, Tai Po, N.T., Hong Kong.
電話　Tel: 2150 2100
傳真　Fax: 2407 3062
電郵　E-mail: info@suplogistics.com.hk

承印者　Printer
中華商務彩色印刷有限公司　C & C Offset Printing Co., Ltd.

出版日期　Publishing Date
二〇一二年九月第一版印刷　First Edition printed in September 2012
二〇一九年五月第四版印刷　Fourth Edition printed in May 2019

版權所有 • 不准翻印　All rights reserved. Copyright © 2012 Wan Li Book Company Limited
ISBN 978-962-14-5901-5

Published in Hong Kong

萬里機構

萬里 Facebook